Stars: A Very Short Introduction

VERY SHORT INTRODUCTIONS are for anyone wanting a stimulating and accessible way into a new subject. They are written by experts, and have been translated into more than 45 different languages.

The series began in 1995, and now covers a wide variety of topics in every discipline. The VSI library now contains over 500 volumes—a Very Short Introduction to everything from Psychology and Philosophy of Science to American History and Relativity—and continues to grow in every subject area.

Titles in the series include the following:

AFRICAN HISTORY John Parker and Richard Rathbone
AGEING Nancy A. Pachana
AGNOSTICISM Robin Le Poidevin
AGRICULTURE Paul Brassley and Richard Soffe
ALEXANDER THE GREAT Hugh Bowden
ALGEBRA Peter M. Higgins
AMERICAN HISTORY Paul S. Boyer
AMERICAN IMMIGRATION David A. Gerber
AMERICAN LEGAL HISTORY G. Edward White
AMERICAN POLITICAL HISTORY Donald Critchlow
AMERICAN POLITICAL PARTIES AND ELECTIONS L. Sandy Maisel
AMERICAN POLITICS Richard M. Valelly
THE AMERICAN PRESIDENCY Charles O. Jones
AMERICAN SLAVERY Heather Andrea Williams
THE AMERICAN WEST Stephen Aron
AMERICAN WOMEN'S HISTORY Susan Ware
ANAESTHESIA Aidan O'Donnell
ANARCHISM Colin Ward
ANCIENT EGYPT Ian Shaw
ANCIENT GREECE Paul Cartledge
THE ANCIENT NEAR EAST Amanda H. Podany
ANCIENT PHILOSOPHY Julia Annas
ANCIENT WARFARE Harry Sidebottom
ANGLICANISM Mark Chapman
THE ANGLO-SAXON AGE John Blair
ANIMAL BEHAVIOUR Tristram D. Wyatt
ANIMAL RIGHTS David DeGrazia
ANXIETY Daniel Freeman and Jason Freeman
ARCHAEOLOGY Paul Bahn
ARISTOTLE Jonathan Barnes
ART HISTORY Dana Arnold
ART THEORY Cynthia Freeland
ASTROPHYSICS James Binney
ATHEISM Julian Baggini
THE ATMOSPHERE Paul I. Palmer
AUGUSTINE Henry Chadwick
THE AZTECS Davíd Carrasco
BABYLONIA Trevor Bryce
BACTERIA Sebastian G. B. Amyes
BANKING John Goddard and John O. S. Wilson
BARTHES Jonathan Culler
BEAUTY Roger Scruton
THE BIBLE John Riches
BLACK HOLES Katherine Blundell
BLOOD Chris Cooper
THE BODY Chris Shilling
THE BOOK OF MORMON Terryl Givens
BORDERS Alexander C. Diener and Joshua Hagen
THE BRAIN Michael O'Shea
THE BRICS Andrew F. Cooper
BRITISH POLITICS Anthony Wright

BUDDHA Michael Carrithers
BUDDHISM Damien Keown
BUDDHIST ETHICS Damien Keown
BYZANTIUM Peter Sarris
CANCER Nicholas James
CAPITALISM James Fulcher
CATHOLICISM Gerald O'Collins
CAUSATION Stephen Mumford and Rani Lill Anjum
THE CELL Terence Allen and Graham Cowling
THE CELTS Barry Cunliffe
CHEMISTRY Peter Atkins
CHILD PSYCHOLOGY Usha Goswami
CHINESE LITERATURE Sabina Knight
CHOICE THEORY Michael Allingham
CHRISTIAN ART Beth Williamson
CHRISTIAN ETHICS D. Stephen Long
CHRISTIANITY Linda Woodhead
CIRCADIAN RHYTHMS Russell Foster and Leon Kreitzman
CITIZENSHIP Richard Bellamy
CLASSICAL MYTHOLOGY Helen Morales
CLASSICS Mary Beard and John Henderson
CLIMATE Mark Maslin
CLIMATE CHANGE Mark Maslin
COGNITIVE NEUROSCIENCE Richard Passingham
THE COLD WAR Robert McMahon
COLONIAL AMERICA Alan Taylor
COLONIAL LATIN AMERICAN LITERATURE Rolena Adorno
COMBINATORICS Robin Wilson
COMMUNISM Leslie Holmes
COMPLEXITY John H. Holland
THE COMPUTER Darrel Ince
COMPUTER SCIENCE Subrata Dasgupta
CONFUCIANISM Daniel K. Gardner
CONSCIOUSNESS Susan Blackmore
CONTEMPORARY ART Julian Stallabrass
CONTEMPORARY FICTION Robert Eaglestone
CONTINENTAL PHILOSOPHY Simon Critchley
CORAL REEFS Charles Sheppard
CORPORATE SOCIAL RESPONSIBILITY Jeremy Moon
COSMOLOGY Peter Coles
CRIMINAL JUSTICE Julian V. Roberts
CRITICAL THEORY Stephen Eric Bronner
THE CRUSADES Christopher Tyerman
CRYSTALLOGRAPHY A. M. Glazer
DADA AND SURREALISM David Hopkins
DANTE Peter Hainsworth and David Robey
DARWIN Jonathan Howard
THE DEAD SEA SCROLLS Timothy Lim
DECOLONIZATION Dane Kennedy
DEMOCRACY Bernard Crick
DEPRESSION Jan Scott and Mary Jane Tacchi
DERRIDA Simon Glendinning
DESIGN John Heskett
DEVELOPMENTAL BIOLOGY Lewis Wolpert
DIASPORA Kevin Kenny
DINOSAURS David Norman
DREAMING J. Allan Hobson
DRUGS Les Iversen
DRUIDS Barry Cunliffe
THE EARTH Martin Redfern
EARTH SYSTEM SCIENCE Tim Lenton
ECONOMICS Partha Dasgupta
EGYPTIAN MYTH Geraldine Pinch
EIGHTEENTH-CENTURY BRITAIN Paul Langford
THE ELEMENTS Philip Ball
EMOTION Dylan Evans
EMPIRE Stephen Howe
ENGLISH LITERATURE Jonathan Bate
THE ENLIGHTENMENT John Robertson
ENVIRONMENTAL ECONOMICS Stephen Smith
ENVIRONMENTAL POLITICS Andrew Dobson
EPICUREANISM Catherine Wilson
EPIDEMIOLOGY Rodolfo Saracci
ETHICS Simon Blackburn
THE ETRUSCANS Christopher Smith
EUGENICS Philippa Levine
THE EUROPEAN UNION John Pinder and Simon Usherwood
EVOLUTION Brian and Deborah Charlesworth

EXISTENTIALISM Thomas Flynn
THE EYE Michael Land
FAMILY LAW Jonathan Herring
FASCISM Kevin Passmore
FEMINISM Margaret Walters
FILM Michael Wood
FILM MUSIC Kathryn Kalinak
THE FIRST WORLD WAR Michael Howard
FOOD John Krebs
FORENSIC PSYCHOLOGY David Canter
FORENSIC SCIENCE Jim Fraser
FORESTS Jaboury Ghazoul
FOSSILS Keith Thomson
FOUCAULT Gary Gutting
FREE SPEECH Nigel Warburton
FREE WILL Thomas Pink
FREUD Anthony Storr
FUNDAMENTALISM Malise Ruthven
FUNGI Nicholas P. Money
THE FUTURE Jennifer M. Gidley
GALAXIES John Gribbin
GALILEO Stillman Drake
GAME THEORY Ken Binmore
GANDHI Bhikhu Parekh
GEOGRAPHY John Matthews and David Herbert
GEOPOLITICS Klaus Dodds
GERMAN LITERATURE Nicholas Boyle
GERMAN PHILOSOPHY Andrew Bowie
GLOBAL CATASTROPHES Bill McGuire
GLOBAL ECONOMIC HISTORY Robert C. Allen
GLOBALIZATION Manfred Steger
GOD John Bowker
GRAVITY Timothy Clifton
THE GREAT DEPRESSION AND THE NEW DEAL Eric Rauchway
HABERMAS James Gordon Finlayson
THE HEBREW BIBLE AS LITERATURE Tod Linafelt
HEGEL Peter Singer
HERODOTUS Jennifer T. Roberts
HIEROGLYPHS Penelope Wilson
HINDUISM Kim Knott
HISTORY John H. Arnold
THE HISTORY OF ASTRONOMY Michael Hoskin
THE HISTORY OF CHEMISTRY William H. Brock
THE HISTORY OF LIFE Michael Benton
THE HISTORY OF MATHEMATICS Jacqueline Stedall
THE HISTORY OF MEDICINE William Bynum
THE HISTORY OF TIME Leofranc Holford-Strevens
HIV AND AIDS Alan Whiteside
HOLLYWOOD Peter Decherney
HUMAN ANATOMY Leslie Klenerman
HUMAN EVOLUTION Bernard Wood
HUMAN RIGHTS Andrew Clapham
THE ICE AGE Jamie Woodward
IDEOLOGY Michael Freeden
INDIAN PHILOSOPHY Sue Hamilton
THE INDUSTRIAL REVOLUTION Robert C. Allen
INFECTIOUS DISEASE Marta L. Wayne and Benjamin M. Bolker
INFINITY Ian Stewart
INFORMATION Luciano Floridi
INNOVATION Mark Dodgson and David Gann
INTELLIGENCE Ian J. Deary
INTERNATIONAL MIGRATION Khalid Koser
INTERNATIONAL RELATIONS Paul Wilkinson
IRAN Ali M. Ansari
ISLAM Malise Ruthven
ISLAMIC HISTORY Adam Silverstein
ISOTOPES Rob Ellam
ITALIAN LITERATURE Peter Hainsworth and David Robey
JESUS Richard Bauckham
JOURNALISM Ian Hargreaves
JUDAISM Norman Solomon
JUNG Anthony Stevens
KABBALAH Joseph Dan
KANT Roger Scruton
KNOWLEDGE Jennifer Nagel
THE KORAN Michael Cook
LATE ANTIQUITY Gillian Clark
LAW Raymond Wacks
THE LAWS OF THERMODYNAMICS Peter Atkins
LEADERSHIP Keith Grint
LEARNING Mark Haselgrove

LEIBNIZ Maria Rosa Antognazza
LIBERALISM Michael Freeden
LIGHT Ian Walmsley
LINGUISTICS Peter Matthews
LITERARY THEORY Jonathan Culler
LOCKE John Dunn
LOGIC Graham Priest
MACHIAVELLI Quentin Skinner
MAGIC Owen Davies
MAGNA CARTA Nicholas Vincent
MAGNETISM Stephen Blundell
MARINE BIOLOGY Philip V. Mladenov
MARTIN LUTHER Scott H. Hendrix
MARTYRDOM Jolyon Mitchell
MARX Peter Singer
MATERIALS Christopher Hall
MATHEMATICS Timothy Gowers
THE MEANING OF LIFE Terry Eagleton
MEASUREMENT David Hand
MEDICAL ETHICS Tony Hope
MEDIEVAL BRITAIN John Gillingham and Ralph A. Griffiths
MEDIEVAL LITERATURE Elaine Treharne
MEDIEVAL PHILOSOPHY John Marenbon
MEMORY Jonathan K. Foster
METAPHYSICS Stephen Mumford
MICROBIOLOGY Nicholas P. Money
MICROECONOMICS Avinash Dixit
MICROSCOPY Terence Allen
THE MIDDLE AGES Miri Rubin
MILITARY JUSTICE Eugene R. Fidell
MINERALS David Vaughan
MODERN ART David Cottington
MODERN CHINA Rana Mitter
MODERN FRANCE Vanessa R. Schwartz
MODERN IRELAND Senia Pašeta
MODERN ITALY Anna Cento Bull
MODERN JAPAN Christopher Goto-Jones
MODERNISM Christopher Butler
MOLECULAR BIOLOGY Aysha Divan and Janice A. Royds
MOLECULES Philip Ball
MOONS David A. Rothery
MOUNTAINS Martin F. Price
MUHAMMAD Jonathan A. C. Brown
MUSIC Nicholas Cook
MYTH Robert A. Segal
THE NAPOLEONIC WARS Mike Rapport
NELSON MANDELA Elleke Boehmer
NEOLIBERALISM Manfred Steger and Ravi Roy
NEWTON Robert Iliffe
NIETZSCHE Michael Tanner
NINETEENTH-CENTURY BRITAIN Christopher Harvie and H. C. G. Matthew
NORTH AMERICAN INDIANS Theda Perdue and Michael D. Green
NORTHERN IRELAND Marc Mulholland
NOTHING Frank Close
NUCLEAR PHYSICS Frank Close
NUMBERS Peter M. Higgins
NUTRITION David A. Bender
THE OLD TESTAMENT Michael D. Coogan
ORGANIC CHEMISTRY Graham Patrick
THE PALESTINIAN-ISRAELI CONFLICT Martin Bunton
PANDEMICS Christian W. McMillen
PARTICLE PHYSICS Frank Close
THE PERIODIC TABLE Eric R. Scerri
PHILOSOPHY Edward Craig
PHILOSOPHY IN THE ISLAMIC WORLD Peter Adamson
PHILOSOPHY OF LAW Raymond Wacks
PHILOSOPHY OF SCIENCE Samir Okasha
PHOTOGRAPHY Steve Edwards
PHYSICAL CHEMISTRY Peter Atkins
PILGRIMAGE Ian Reader
PLAGUE Paul Slack
PLANETS David A. Rothery
PLANTS Timothy Walker
PLATE TECTONICS Peter Molnar
PLATO Julia Annas
POLITICAL PHILOSOPHY David Miller
POLITICS Kenneth Minogue
POPULISM Cas Mudde and Cristóbal Rovira Kaltwasser
POSTCOLONIALISM Robert Young

POSTMODERNISM Christopher Butler
POSTSTRUCTURALISM
Catherine Belsey
PREHISTORY Chris Gosden
PRESOCRATIC PHILOSOPHY
Catherine Osborne
PRIVACY Raymond Wacks
PSYCHIATRY Tom Burns
PSYCHOLOGY Gillian Butler and
Freda McManus
PSYCHOTHERAPY Tom Burns and
Eva Burns-Lundgren
PUBLIC ADMINISTRATION
Stella Z. Theodoulou and Ravi K. Roy
PUBLIC HEALTH Virginia Berridge
QUANTUM THEORY
John Polkinghorne
RACISM Ali Rattansi
REALITY Jan Westerhoff
THE REFORMATION Peter Marshall
RELATIVITY Russell Stannard
RELIGION IN AMERICA Timothy Beal
THE RENAISSANCE Jerry Brotton
RENAISSANCE ART
Geraldine A. Johnson
REVOLUTIONS Jack A. Goldstone
RHETORIC Richard Toye
RISK Baruch Fischhoff and John Kadvany
RITUAL Barry Stephenson
RIVERS Nick Middleton
ROBOTICS Alan Winfield
ROMAN BRITAIN Peter Salway
THE ROMAN EMPIRE Christopher Kelly
THE ROMAN REPUBLIC
David M. Gwynn
RUSSIAN HISTORY Geoffrey Hosking
RUSSIAN LITERATURE Catriona Kelly
THE RUSSIAN REVOLUTION
S. A. Smith
SAVANNAS Peter A. Furley
SCHIZOPHRENIA Chris Frith and
Eve Johnstone
SCIENCE AND RELIGION
Thomas Dixon
THE SCIENTIFIC REVOLUTION
Lawrence M. Principe
SCOTLAND Rab Houston
SEXUALITY Véronique Mottier
SHAKESPEARE'S COMEDIES Bart van Es
SIKHISM Eleanor Nesbitt
SLEEP Steven W. Lockley and
Russell G. Foster
SOCIAL AND CULTURAL
ANTHROPOLOGY
John Monaghan and Peter Just
SOCIAL PSYCHOLOGY Richard J. Crisp
SOCIAL WORK Sally Holland and
Jonathan Scourfield
SOCIALISM Michael Newman
SOCIOLOGY Steve Bruce
SOCRATES C. C. W. Taylor
SOUND Mike Goldsmith
THE SOVIET UNION Stephen Lovell
THE SPANISH CIVIL WAR
Helen Graham
SPANISH LITERATURE Jo Labanyi
SPORT Mike Cronin
STARS Andrew King
STATISTICS David J. Hand
STUART BRITAIN John Morrill
SYMMETRY Ian Stewart
TAXATION Stephen Smith
TELESCOPES Geoff Cottrell
TERRORISM Charles Townshend
THEOLOGY David F. Ford
TIBETAN BUDDHISM
Matthew T. Kapstein
THE TROJAN WAR Eric H. Cline
THE TUDORS John Guy
TWENTIETH-CENTURY BRITAIN
Kenneth O. Morgan
THE UNITED NATIONS
Jussi M. Hanhimäki
THE U.S. CONGRESS Donald A. Ritchie
THE U.S. SUPREME COURT
Linda Greenhouse
THE VIKINGS Julian Richards
VIRUSES Dorothy H. Crawford
VOLTAIRE Nicholas Cronk
WAR AND TECHNOLOGY
Alex Roland
WATER John Finney
WILLIAM SHAKESPEARE
Stanley Wells
WITCHCRAFT Malcolm Gaskill
THE WORLD TRADE
ORGANIZATION Amrita Narlikar
WORLD WAR II Gerhard L. Weinberg

Andrew King

STARS

A Very Short Introduction

Great Clarendon Street, Oxford, OX2 6DP,
United Kingdom

Oxford University Press is a department of the University of Oxford.
It furthers the University's objective of excellence in research, scholarship,
and education by publishing worldwide. Oxford is a registered trade mark of
Oxford University Press in the UK and in certain other countries

First Edition published in 2012

Impression: 11

British Library Cataloguing in Publication Data
Data available

Library of Congress Cataloging in Publication Data
Data available

ISBN 978–0–19–960292–6

Printed and bound by
CPI Group (UK) Ltd, Croydon, CR0 4YY

Contents

Preface

Every atom of our bodies has been part of a star, and every informed person should know something of how the stars evolve. This book aims to explain how the laws of physics force stars to evolve, driving them through successive stages of maturity before their inevitable and sometimes spectacular deaths, to end as remnants such as black holes.

The book should be accessible to anyone with some recollection of school science. The reader should, for example, be prepared to accept that the area of a sphere is related to the square of its radius, and have some familiarity with notions such as pressure and density. Simple equations occasionally appear, but only when they are (for those who know a little mathematics) simpler to understand than the verbal descriptions I also give. I hope the book will be useful to anyone taking a first course in astronomy, such as school science modules, specialist university courses, and 'astronomy/physics for poets' courses in the US.

I thank Rob Colls, a fellow sufferer in the university sports hall, who urged me to write this book, and offered encouragement and advice. Chris Nixon produced several of the figures, and Lisa Brant got the final version into a publishable format.

This book was written partly in rural Leicestershire, and partly in Amsterdam. I thank Nicole for making me a home in both places.

List of illustrations

Chapter 1
Science and the stars

What are stars made of?

In 1835, the philosopher Auguste Comte, the founder of sociology, and whose notions of order and progress are inscribed on the flag of Brazil, wrote:

> there is no conceivable means by which we shall one day determine the chemical composition of the stars.
>
> (*Cours de Philosophie Positive*)

Shortly after his death in 1857, science proved Comte's categorical statement wrong: the stars contain all the elements we find on Earth. And we now know that every atom in our bodies was once part of a star, or even several stars. In a few years, the nature of the stars moved from a mystery that it was pointless – or even forbidden – to speculate on, to the basis of how we understand the Universe and our place in it. From the preserve of dogma and mysticism, the workings of the stars became a legitimate subject of quantitative scientific enquiry.

Physics caused this revolution. Comte's error was to assume that the only possible way of discovering what anything was made of was to get hold of a sample of it. But science leads you to thoughts you could not have imagined. We can deduce the chemical composition of something just by examining the light it emits when it is hot. Although there had been many clues, the clinching discovery belongs to two German physicists, Gustav Kirchhoff and Robert Bunsen. They knew that flames from chemicals involving different elements burn with different colours. Copper compounds produce bluish flames, sodium ones intense yellow, and so on. But this did not really tie down the nature of an unknown sample, because the connection between the chemical element and the colour is not unique – arsenic and lead compounds also produce blue flames, for example.

Crucially, Kirchhoff realized that you needed to know in some precise quantitative way just *how* red or blue the light from the heated sample was. He spread out the light by passing it through a prism. Isaac Newton first did this with sunlight in the 17th century, and found that it was composed of all the colours of the rainbow, from red to blue-violet. Joseph von Fraunhofer, a German optician, went further in 1814, showing that this *spectrum* of the Sun's light was crossed by hundreds of dark lines. These always appeared in the same places, and we now know that these positions correspond to the wavelength of the light involved – longer for red light, shorter for blue light. Spreading out the light in this way allowed Kirchhoff and Bunsen to see that the light given off by a heated sample of any given element produced a precise pattern of *bright* lines in the spectrum. This pattern gave a unique fingerprint of the element: if you saw this pattern of bright lines at these specific wavelengths, you knew that this element was present. Bunsen's contribution was to invent the famous burner that you probably remember from school chemistry, for heating the samples. This was vital, because its flame did not emit any visible lines of its own which would have messed up the identification of the characteristic line spectra of each element.

It did not take Kirchhoff and Bunsen long to realize that they could use their instrument – now called a *spectrograph* – to work out the composition not just of laboratory samples, but also of distant things, if they were hot enough. One evening, they could see the flames of a huge fire in Mannheim from their laboratory in Heidelberg, ten miles away. Their spectrograph detected lines of barium and strontium in them. From here, it was a short but momentous step to using their spectrograph on sunlight. Of course, they again found Fraunhofer's dark lines, rather than the bright ones they had seen from heated samples. But now they noticed that the positions of many of these dark lines – and so their wavelengths – corresponded exactly to the positions of bright lines they had seen from heated laboratory samples. This was obviously no coincidence. Fraunhofer's dark lines corresponded to light being *removed* from the Sun's spectrum at the same wavelengths that these elements *added* light when heated in the laboratory.

Kirchhoff quickly realized what this meant: a given chemical element emitted light at various characteristic wavelengths if it was hotter than its surroundings, but *absorbed* light at these same wavelengths if it was cooler. He imagined two samples of the same element, heated to different temperatures and enclosed in a box. All experience said that the two samples would settle at the same temperature: the hotter one would cool, and the cooler one heat up, until this happened. But their only way of communicating any information about their temperatures was by the light they emitted and absorbed from each other – so the cooler sample must absorb light very efficiently, at exactly the characteristic pattern of wavelengths that the hotter one emitted (see Figure 1). Fraunhofer's dark lines in the solar spectrum meant that the outer layers of the Sun were cooler than those further in, so that the light at those wavelengths was absorbed and never reached the Earth.

All this said nothing about the origin of the characteristic pattern of wavelengths for each element. We now know that these lines tell us about the structure of the atoms of each element, and we will say more about this later. But even without this knowledge, only fully appreciated in the 1920s, the shift in thinking was

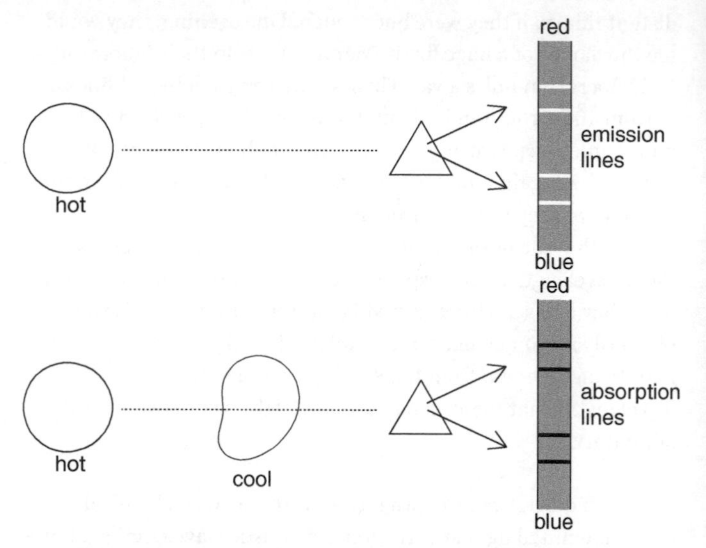

1. Emission and absorption lines. Top: the light from a hot sample is spread out by passing it through a prism. It emits a pattern of bright lines (emission lines) characteristic of the elements present in it. Bottom: if the same sample is illuminated from behind by a hotter source, the spectrum lines appear darker than the rest of the spectrum (absorption lines). The outer layers of a star are cooler than its interior, so we generally see an absorption line spectrum from it

profound. Science now had a definite physical picture of the Sun as a ball of hot gas, cooler at its surface than in the interior, acting just as any hot body would (see Figure 2). The centre of a burning lump of coal is hotter than its surface, just as your bed is warmer under the covers than outside them.

The rest of this book will develop this picture, and lead us very far. But first we need to understand what basic facts astronomers know about stars, and how they know them. If you already know these things, or are prepared to take them on trust, you can skip to the last section of this chapter.

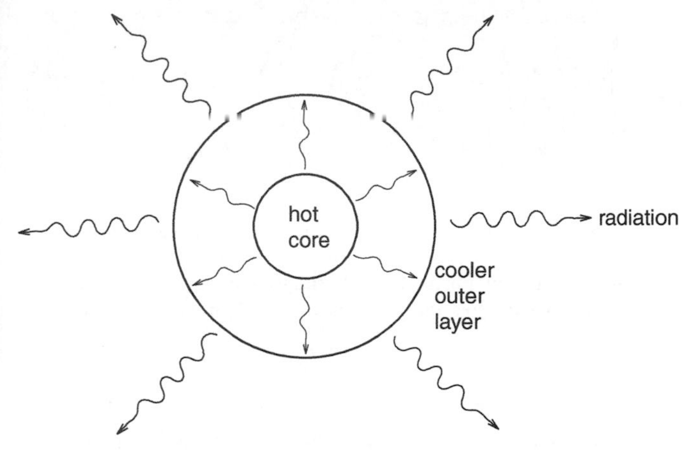

2. A star loses heat into space from its surface. Its temperature is highest in the centre and lowest at the surface

How heavy, how hot, how bright, how big?

Attach a spectrograph to a telescope – something first done by the wealthy English amateurs William and Margaret Huggins – and you have the key to understanding the stars. Amazingly, it can tell us their masses. The way to this is through the Doppler effect, which affects all wavelike phenomena such as sound waves as well as light (see Figure 3). This effect allows us to use spectra to work out speeds. If the object emitting the waves is moving away from us, each successive wave has slightly further to travel to reach us than the one before. So the wave crests arrive slightly further apart in time than before – the wavelength λ we measure is *longer*, and the wave's frequency v is lower. (The wave speed $c = \lambda v$ remains fixed.) If these are light waves, this longer wavelength means a slight shift towards the red end of the spectrum, a redshift. If they are sound waves, the longer wavelength means a lower pitch (big instruments make lower notes than small ones: think of a double bass and a violin). Conversely, if the source of the waves moves towards us rather than away, successive wave crests arrive slightly

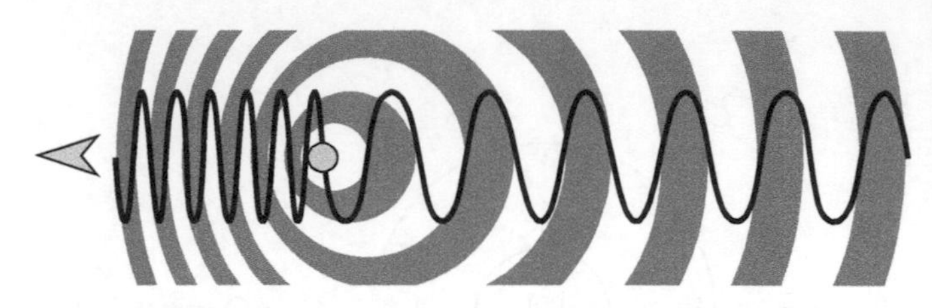

3. The Doppler effect: as a radiation source moves towards the viewer, the wave crests arrive closer together in time, so the measured wavelength is shorter (blueshifted). As the source moves away, the wavecrests arrive further apart in time, and the light is redshifted

early, and the wavelength we measure is *shorter*: a blueshift for light, and a higher pitch for sound. You can hear the Doppler effect very easily as a police car speeds past you with its siren sounding – the pitch suddenly drops as it passes you and starts to move away. (The gradual rise in pitch as it approaches you is less noticeable, as the sound is fainter at the start and the rise is slow.)

Applied to the spectra of stars, the Doppler effect tells us if the star is moving away or towards us. There is no Doppler shift for sideways motion (in the plane of the sky), so the redshift or blueshift just tells us the velocity along the line of sight. The relation between this velocity and the Doppler shift is simple: the fractional change in the wavelength is the ratio of the velocity v to the speed of light c. Mathematically, this is expressed as $\Delta\lambda/\lambda = v/c$, where $\Delta\lambda$ is the change of wavelength. So if we measure all the lines from a source to have wavelengths 1% longer than the same lines in the laboratory, the star producing these lines (usually dark or absorption lines) must be moving away from us at 1% of the speed of light, or about 3,000 kilometres per second. Usually the Doppler shifts and velocities are much smaller than this, so it is important to be able to measure wavelengths to high accuracy.

How do we get the masses of stars from all this? Astronomers know of many pairs of stars which are orbiting each other under

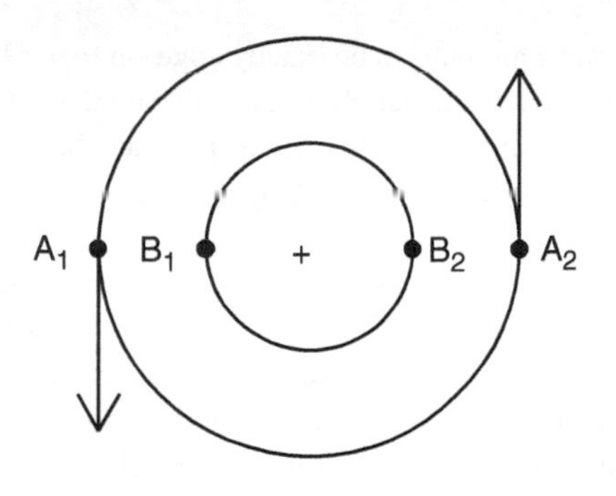

4. Spectroscopic binary stars. Stars 1 and 2 orbit each other, in this case in circular orbits, under their mutual gravitational attraction. When star 1 is at A_1 (and star 2 at A_2), the observer measures a Doppler blueshift from star 1. At B_1 (and B_2), she measures an equal and opposite redshift from this star. Observing the spectrum of a binary system over time in this way gives information on the speed of one or both stars, and so allows astronomers to estimate the masses of the stars

their mutual gravitational attraction. The laws of motion tell us that these orbits lie in the same plane, at a fixed angle to our line of sight. As the stars move around in their orbits, they move alternately slightly away and slightly towards us – their component velocities along our line of sight oscillate backwards and forwards. So spectral lines from each of them move back and forth across the spectrum about their average, or mean, wavelengths (see Figure 4). These mean wavelengths may themselves be shifted from the laboratory value, which tells us that the whole binary system is itself moving bodily away or towards us, like a revolving carousel carried on the back of a truck. In the simplest case, the biggest deviations from this mean value (the maximum red – and blue – shifts) are equal, and tell us that the two stars are moving in circular orbits about each other and their centre of mass. If the plane of this orbit happens to be exactly edge-on to our line of sight, this deviation immediately gives us its circular velocity around the centre of mass. If we are exceptionally lucky, we can do this for both stars in the binary system.

Physically, we know that these motions are governed purely by gravity: the stars avoid falling in towards each other because of their circular motions. The gravitational force between two bodies is proportional to the product of their masses and decreases as the square of the distance between them. Given this, we now have enough information to work out what the masses of each of the stars must be.

This method, and elaborations of it used to deal with more complicated cases, forms the basis of our knowledge of stellar masses, with one exception: in Chapter 2, we shall see that we know the mass of the Sun very accurately because we can observe its gravitational effect on all the planets and smaller bodies in the Solar System. It turns out that stellar masses range from about ten times less than that of the Sun, to something like a hundred times larger than it. Astronomers use the Sun as a mass unit, and so refer to this range as running from 0.1 to 100 solar masses.

If stars are hot bright balls of gas, the obvious questions are how hot, and how bright? How hot is relatively easy: physics tells us that hot opaque bodies emit light with a characteristic smooth distribution over wavelength depending only on their surface temperature: the hotter the star, the more blue light it emits relative to the red, making it look whiter. Simply filtering the light received by a telescope – blocking the red and then the blue light – to view a star in its blue light and red light separately already gives us a good idea of its surface temperature. These temperatures range from about 2,500 degrees up to 30,000 or more. The 'degrees' here are on the Kelvin or absolute scale. We will see in Chapter 2 that these are almost the same as the familiar Celsius scale.

The question 'how bright' is rather harder, as we need to know how far away a star is to decide this. The radiation flux from a spherical body like a star decreases as the *square* of its distance from us. So a star looks four times fainter at twice the distance, nine times fainter at three times, and so on. The first stellar distances were found by what are called parallaxes (see Figure 5). The principle is very simple. Close one eye, and with your arm half extended, put your forefinger over some fixed object like the edge of a door, or a window. Now open your other eye and close the first, and your finger appears to move relative to the door or window behind it. Now repeat the experiment, but this time with your arm fully extended. Your finger still appears to move, but less – about one-half as far. A stellar parallax is the same thing with three substitutions: your finger is a nearby star, and the window or door edge is the background of distant stars. Your two eyes – two viewing points of the nearby star against the distant stars – are chosen as the positions of the Earth in its orbit at two dates six months apart. This means that the distance between these two points is just the diameter of the Earth's orbit around the Sun. Nearby stars appear to move slightly relative to distant ones when viewed from these two positions (see Figure 5). Measuring the angle by which they seem to move, and knowing the size of the Earth's orbit, we have enough information to work out how far away the nearest stars are. This method clearly does not work well for more distant stars, as the angle that they appear to move becomes very small. However, the stellar parallax method gives us a first step in constructing a self-consistent system of distances. We shall say much more about this in Chapter 7.

Given a distance, we can use the radiation flux we collect from the star to find the total amount of light energy it emits – its *luminosity*. Our telescope intercepts only a tiny fraction of this light, given by the ratio of the area A of the telescope mirror and the vastly larger area of a sphere centred on the star, and passing through the Earth, that is by 4π times the square of its distance D. Correcting the amount of radiation energy received at the

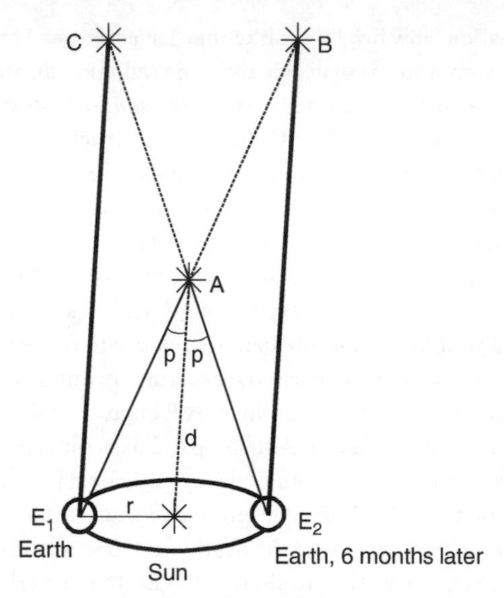

5. Stellar parallax. Nearby star A is first observed from Earth at point E_1 in its orbit around the Sun, and then six months later when the Earth is at the opposite point E_2 of its orbit. Star A appears to have moved slightly in the sky relative to distant stars such as B and C, giving the angle *p*. This angle is given by the ratio of the radius *r* of the Earth's orbit to the distance *d*. Since *r* is known, we can work out the distance *d* to the star

telescope by this huge ratio $4\pi D^2/A$ gives us its luminosity. Now we also get the radius of the star, because the luminosity L of a hot body is proportional to its total surface area (here $4\pi R^2$) times its temperature T to the power four, so that $L \propto 4\pi R^2 T^4$. For a star of 10,000 degrees, this fourth power is 10^{16}, or one followed by sixteen zeroes (or one hundred thousand billion, if you prefer). So if we know the total luminosity and the temperature of a star, we can work out its surface area, which is just 4π times the square of its radius R. Astronomers now know the radii of many stars. They turn out to range from about one-tenth of the Sun's radius, to sizes similar to the distance from the Earth to the Sun, about 200 times the Sun's radius.

Perhaps surprisingly given the flying start the subject had from spectroscopy, the precise composition of the stars, the proportions of each element by mass in a star, took rather longer to settle than masses, luminosities, and temperatures. The problem is that each element only reveals itself readily by spectral lines at a favoured range of temperatures. Hydrogen is very prominent in the spectra of stars with surface temperatures around 10,000 degrees, but difficult to detect outside this range, and so on. The mere absence of a spectral line does not allow us to say that the element producing it is not present in the star. The reason for this enigmatic behaviour goes back to the structure of the atom, which we will describe in Chapter 3. So for a long time, astronomers had little idea what were the most common elements in stars. The first person to use a modern theory to work out when various spectral lines would appear was Cecilia Payne, a young English-born PhD student at Radcliffe College (now part of Harvard University), in 1925. She found that all the spectroscopic data were consistent with the idea that hydrogen was by far the most abundant element in all but a very few stars, with helium a distant second. This totally contradicted the orthodox view at the time, which was that stars probably had a similar composition to the Earth. Payne's PhD thesis was reviewed by Henry Norris Russell of Princeton, who dissuaded her from publishing her conclusion. This only became the universally accepted view four years later. Cecilia Payne went on to become the first female full professor at the Harvard Observatory.

Stars are simple

Early 20th-century astronomers did not know what stars were made of, or even (as we shall see) why they shine at all. Despite this, they made considerable progress. They knew the luminosity L and surface temperature T of a reasonable collection of stars. Eventually, the Danish astronomer Ejnar Hertzsprung and the same Henry Norris Russell independently did the obvious thing of plotting a graph of one of these quantities against the other.

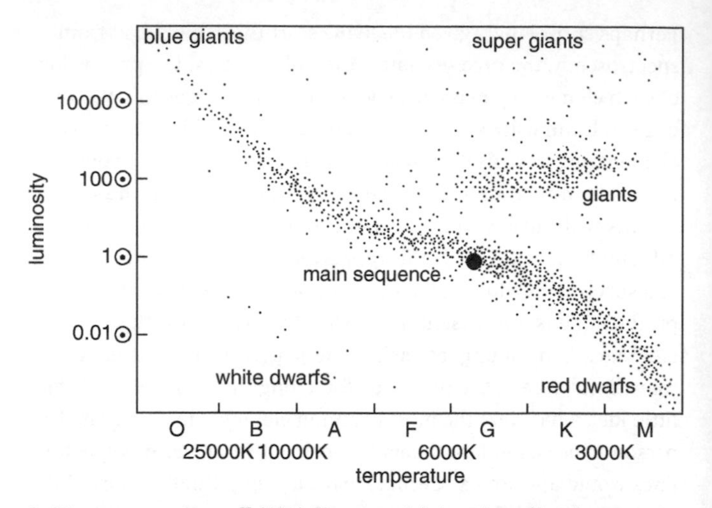

6. Hertzsprung–Russell (HR) diagram: this is a plot of luminosity against surface temperature. Stars do not lie in random positions on this diagram, but mostly in well-defined families. The main sequence is composed of stars burning hydrogen in their cores. Red giants are stars which have ended hydrogen core burning, and have hugely extended envelopes surrounding nuclear shell-burning layers around inert compact cores. White dwarfs are the remnants of giants which have lost their envelopes and shell sources, and consist of the small but very dense remaining degenerate cores. The dot denotes the Sun

For historical reasons, this graph is always drawn with the temperature *decreasing* to the right, opposite to the normal convention. The resulting diagram, known as the Hertzsprung–Russell, or HR, diagram, is the most important figure in stellar astrophysics, and we will use it repeatedly in this book (Figure 6). As you can see, stars are not scattered randomly over this diagram, but are restricted to various well-defined families. The most populous family, running diagonally from top left to lower right, is called the main sequence. This name comes from an old theory of stellar evolution which is now discredited. In reality, stars do not move along this 'sequence'. The slope of the main sequence tells us that the luminosities of stars on it are very sensitive to their surface temperatures – luminosity goes roughly like the seventh power of temperature (i.e. $L \propto T^7$), so that doubling the temperature implies luminosity more than a hundred times larger. Branching off the main sequence towards the upper right is a less numerous branch of cool but luminous stars. Now we already know that the luminosity of a star is proportional to its area multiplied by the temperature to the fourth power (i.e. $L \propto 4\pi R^2 T^4$): the only way that these cool stars can be so luminous is if they have very large areas, or equivalently, large radii. These are the largest stars, with radii comparable with the Earth–Sun distance of about 200 solar radii: we call these red giants. Note that there is no reason at all to imagine that these giants have greater *masses* than other stars of smaller radius: we will see later that they do not; they are simply more diffuse, spreading a similar mass over a much larger volume. Finally, there is a small group of hot but faint stars at the bottom left: these must have very small radii, comparable to the Earth's own radius, and we call these white dwarfs. Again, we will find that these stars too have similar masses to other stars, and so are very dense.

At one blow, the HR diagram tells us that stars are simple objects – we can already get a good description of most of them simply by specifying just two numbers: their luminosity and surface temperature. But the diagram challenges us to use physical laws to explain why a hot ball of gas generally settles on just one of the three families: main sequence, giant, and white dwarf. Are these stars physically distinct in some way, or do the various families on the HR diagram represent different stages in the evolution of stars?

Chapter 2
How the Sun survives

Gravity and the Sun

The Sun is our nearest star. Human experience already tells us a lot about it, and so about stars in general. Two things are so obvious that we rarely question them: the Sun radiates vast amounts of energy into space – we on Earth benefit from intercepting just a tiny fraction of this light – and the Sun's gravity keeps the Earth in an orbit we complete once a year. These two facts are central to our existence. With only a few small exceptions, all plant and animal life on Earth depends directly on the Sun. The energy passed down the food chain starts as sunlight captured by photosynthesis in plants. When we burn fuel, we are releasing stored energy from the same ultimate source. Water falls and powers electric generators only because sunlight lifts it high into the atmosphere and makes rain and snow. And life is possible only because the Sun warms the Earth, and the atmosphere it creates keeps temperatures nearly constant and spreads water over the Earth.

But this energy input has to be very precise: if our orbit were only a little closer to the Sun (like Venus), the Earth would probably be too hot for life. And if we orbited only a little further away from the Sun (like Mars), the Earth would probably be too cold. Equivalently, if the Sun were just a little hotter or cooler, life would probably never have appeared on the Earth. Things are so delicate that if the Sun changed the amount of energy it shines on the Earth by a relatively small amount, life would probably disappear. So the *stability* of the Sun's output is crucial to life on Earth.

In fact, we know for all sorts of reasons that the Earth has supported life in some form for almost four billion years, almost as long as the age of the oldest minerals on Earth. This must mean that the Sun's output – its luminosity – has remained almost constant over this vast timescale, which is almost one-half of the age of the entire Universe. So we can deduce that the Sun is an extraordinarily stable object. This immediately tells us a lot more about the Sun, things that have to be true for it to remain the unchanging object we know. For example, all the forces acting inside the Sun must balance each other almost perfectly. If they did not, one or more of them would overwhelm the others, and the Sun would change radically from the object that we know into something that we would not recognize, and which could not have supported life on the Earth for so long.

The most obvious force that acts on the Sun is gravity, the attraction that acts between all forms of matter. We experience the gravitational pull of the Earth on our bodies, and on things around us, as *weight*. Compared with the other fundamental forces which we shall meet later in this book, gravity is extremely weak. You may find this a strange remark, if you recall the effort involved in climbing a tall flight of stairs. But think of it this way: our muscles are able to overcome the gravity of the entire Earth pulling upon us, leaving us free to stand and move on its surface. The weakness of gravity is essential to life. If gravity were comparable in strength to the other basic forces, ordinary objects would attract each other fiercely, whereas, for example, the force between your hand and this book is (fortunately) entirely negligible.

So if gravity is so weak, why all the fuss about it? Surprisingly, gravity is by far the most important force in shaping the Universe. Two things about it mean that it dominates all the other forces over large distances. These properties ensure that despite gravity's relative weakness, the gravitational pull of the Sun is what swings the planets around in their orbits. All the planets in the Solar System move entirely according to the dictates of the Sun's gravity. Their motions tell us that the Sun's mass is about 2×10^{30} kilograms, or 2×10^{27} tons. This is about 300,000 times more massive than the Earth.

What are these two properties that make gravity the dominant force at large distances? First, gravity is always attractive and never repulsive. This is quite unlike the force between two electric charges. This *electrostatic* force is strongly attractive if one of the charges is positive and the other negative, but equally strongly repulsive if they are both positive or both negative. So in a mixture of positive and negative charges, there is a powerful tendency for the attractive and repulsive electrical forces to cancel each other out. But what if we have a mixture where the number of positive and negative charges are not equal? Ironically, the electrostatic force then falls victim to its own strength: a body with a net electric charge tends to neutralize that charge, and so make electrostatic interactions extremely weak in practice. As an example, let's imagine a body with more positive charges than negative ones. Because its overall net charge is positive, it strongly repels any nearby positive charges, and attracts negative ones equally strongly. Nothing can slow this process down, precisely because of the strength of the electrostatic force between two charges. Gravity is powerless to affect this process, for example. As these negative charges stick to the body's surface, they act to neutralize the original positive charge.

You might ask where these 'nearby charges' can come from if the body we are interested in is a star – after all, don't astronomical bodies like the Sun sit in regions of space which are remarkably like a complete vacuum? But the Universe appears to be endowed with precisely equal numbers of positively and negatively charged particles. This makes it very hard ever to separate large numbers of positive charges from negative ones, especially as the two separate collections then want to disperse themselves. So to charge a body, you have to pull a few small charges off it, and somehow prevent them returning to neutralize it. Since this is so difficult to arrange, large bodies tend to retain closely equal numbers of positive and negative charges. The electric force between two bodies like this consists of myriad pairs of attractive and repulsive forces that cancel each other out. All this happens because the electric force is so strong – so strong that no other force can stop it moving charges to cancel itself out.

But gravity is different – in its modest way, it always attracts, with a strength gauged by multiplying together the two masses involved. So the gravitational force between two objects with masses 2 and 3 kilograms respectively, is $2 \times 3 = 6$ times stronger than between two objects each having a mass of 1 kilogram. Similarly, the attraction between masses of 10 and 100 kilograms is $10 \times 100 = 1000$ times stronger. Of course, this force is still tiny – so small that only very accurate laboratory experiments can measure it, but we can see that adding more mass will always increase the gravitational attraction. Gravity has a second property that makes it dominant on large scales. Its strength does decrease as the distance between the attracting bodies grows, but this decrease is relatively slow – as the *inverse square* of this distance. This means that if the distance is doubled, the gravitational force decreases by $1/2 \times 1/2 = 1/4$, if it is tripled it decreases by $1/3 \times 1/3 = 1/9$, and so on. Other forces we shall meet later in this book drop off far more rapidly than this inverse square behaviour, and so are negligible except on tiny scales comparable with an atomic nucleus.

So gravity has the unique twin properties of always being attractive, and relatively long range. This means that it always beats all the other forces for large objects, and so is the most important force in astronomy. Crucially for stars, this gives every astronomical body a weight problem: it has to find some way of balancing itself against gravity, or collapse in on itself if it cannot. This is clearly a big problem for the Sun, because of its huge mass. The motions of the planets tell us its gravitational attraction is very powerful even at large distances. But gravity is also acting *inside* the Sun, each part of it attracting all the others. These internal parts of the Sun are far closer together than any of the planets, making the attraction much stronger (by the inverse square effect we discussed above). The Sun is almost perfectly spherical, so the outcome of all the internal gravitational attractions within it must be a force trying to shrink it inwards towards its centre. Even at the surface of the Sun, this force is more than ten times stronger than gravity on the Earth. If you could stand on the Sun's surface, you would feel ten times heavier than you do now, and things would get worse if you could move further in. This makes it obvious that some other force must be resisting gravity's pull on the Sun: if not, it would collapse inwards under its own weight. This collapse would be horribly sudden: the radius of the Sun would decrease significantly in only a hour, and its appearance would change dramatically in that time. These are things that we know for all kinds of reasons simply do not happen, and have almost certainly never happened for the four billion years the Earth has supported life.

Now we have an important question to answer. What force resists gravity's tendency to make the Sun collapse inwards? We know that the Sun is made of gas – something first suggested by the early work of Kirchhoff and Bunsen that we discussed in Chapter 1, and nowadays best appreciated by viewing movies of its heaving surface. One obvious feature of any gas is *pressure*. This suggests that the solar gas might be able to hold itself up against gravity by exerting pressure against its weight.

Although the idea of pressure is familiar from everyday life, we should ask what it really means for the Sun. Pressure in a gas comes about because its constituent particles – molecules, or atoms – are all moving relative to one another. This movement is directly related to the temperature of the gas: the hotter the gas, the faster its particles move. The motions of the gas particles that constitute pressure are not orderly, that is, all streaming in one direction, but chaotic. In fact, the very word *gas* was synthesized by the 17th-century Flemish chemist Johannes van Helmont from the Greek word *chaos* (although, confusingly, it then referred to empty space). This chaotic motion means that the particles collide, producing forces on each other (which are actually electrical). Clearly, the pressure depends on how many particles there are in a given volume of gas, or equivalently by the mass of gas in the same volume (its *density* ρ), and on how fast they are moving – specified by the temperature (T). We can choose to measure the temperature on a scale where zero degrees corresponds to zero motion of the gas particles. This is the so-called absolute temperature scale and amounts to adding 273 degrees on to the Celsius temperature. So 20 degrees Celsius becomes 293 degrees absolute, or 293 K, where K refers to Lord Kelvin, the Scottish physicist who devised the absolute scale. In the kind of gas making up the Sun, the pressure is given simply by multiplying these two numbers together, and then further multiplying by a known constant number (i.e. $P \propto \rho T$). This relation is often called the perfect gas law, and contains both Boyle's law (pressure times volume is constant at fixed temperature for a fixed mass of gas) and Charles' law (volume is proportional to absolute temperature at fixed pressure for a fixed mass of gas).

But pressure by itself is not enough. Imagine that the pressure is the same everywhere in the Sun. Particles collide back and forth, but the force delivered by every collision is always cancelled out by another one in the opposite direction very nearby. This will not help the Sun to support its own weight. To get a *net* pressure force pushing gas away from the centre, we must ensure that this cancellation of forces does not happen. If we can arrange that the pressure *increases* in the direction that the weight pushes, we have a promising candidate for the force that holds up the Sun against its own weight. The Sun will stay in mechanical balance if the pressure increases inwards, in line with the weight of the overlying gas at each depth inside it, just as water pressure increases with depth in the sea. So the pressure in the centre of the Sun must be able to cope with the entire weight of its 2×10^{27} tons, all pulling inwards on itself.

Clearly, the pressure resisting this must be huge, and we should ask what it is that makes the pressure so high. In the reasoning above, we worked out that pressure is given by multiplying density and temperature together. So to make a huge pressure, one or both of the density or the temperature in the centre of the Sun must be extremely high. In fact, we can guess that a high density is unlikely to help – this only means that denser gas is closer to the centre, where gravity is stronger, thus increasing the weight. The origin of gas pressure is the motion of the gas particles, specified by the absolute temperature, and we want the ones in the centre to have higher speeds than those at the surface. The reason the Sun does not collapse is that its central temperature is high.

This central temperature, and the pressure it implies, is the only thing that keeps the Sun at its present size, despite the gravity exerted by its mass. So we can intuitively guess that the central temperature is connected with both the Sun's size and its mass. We can think of this temperature as giving the central gas particles in the Sun average energies that would let them compete with gravity at the Sun's surface, if they could reach it from the centre without colliding with other gas particles further out. For example, they could move in orbits that reach out to distances like that of the surface, that is, the Sun's radius. Bodies orbiting in this way have speeds specified by dividing the Sun's mass M by its radius R and taking the square root ($v = \sqrt{GM/R}$, with G the gravitational constant). So the bigger the mass, the higher the speed, and the bigger the radius, the lower this speed. The central temperature T_* controls these speeds, and so, like them, is also specified by the ratio of the Sun's mass to its radius (but now without the square root, i.e. $T_* \propto v^2 \propto GM/R$).

Now we can make a simple estimate of the temperature T_* in the centre of the Sun from a knowledge of its mass and its radius. The resulting central temperature is very high, in fact about 10 million degrees Kelvin – far higher than the surface temperature T of about 6,000 degrees Kelvin. We will see later that the precise value of the central temperature is very significant.

Energy and evolution

We have argued that the Sun must remain stable for a very long time, simply because of the known age of the Earth. We can be confident that its high central temperature does stabilize the Sun against the immediate effects of gravity, and there is no danger that it will collapse and change violently in the course of an hour. But this is not necessarily enough to ensure a long stable life. The reason is the Sun's most obvious characteristic of all – it shines, losing energy by radiating it into space. We know that this is a huge amount every second. Unless we occupy some weirdly special position, the amount of solar energy reaching the top of the Earth's atmosphere should be typical of what it radiates in every other direction. This amount is about 1.4 kilowatts per square metre. (The average amount we receive at the Earth's *surface* is only one-quarter of this, fortunately for us, because the radiation intercepted by the side of the Earth facing the Sun gets spread over all of its surface by the atmosphere, and the area of a sphere is 4 times that of a flat disc of the same radius r [$4\pi r^2$ and πr^2 respectively].) We can work out what the Sun radiates by multiplying this 1.4 kilowatts per square metre by the area of a sphere centred on the Sun, with radius equal to our distance D from the Sun. This distance is about 150 million kilometres, which is 1.5×10^{11} metres, so the Sun's output works out at $1400 \times 4\pi D^2 = 4\pi \times (1.5 \times 10^{11})^2$ watts, or a staggering 3.8×10^{26} watts.

Since energy is conserved, something must be changing inside the Sun to provide this huge output. The two physicists who independently first studied this problem in the 19th century were Kelvin (whom we encountered earlier) and Hermann von Helmholtz. They were among the first physicists to recognize that energy is always conserved. Helmholtz was driven to this conclusion by his early medical studies of muscle metabolism, through which he convinced himself that the then popular notion of a 'vital force' animating living tissue was nonsense: the energy expended by animals is simply equal to the food energy they consume.

Applied to the Sun, this conservation principle implies that there must be a store of energy which is being depleted as the Sun shines. The Sun has an internal clock – as this energy store is depleted, it must *evolve*. This is our first encounter with a concept – stellar evolution – which will dominate this book. The stars have lives, which are *finite*. So how long do the stars live, especially the Sun? This type of question is familiar: if my business is losing money, how long before it folds? Of course, it all depends on how much money you have stored, or in the Sun's case, how much energy. So we need some idea of the source of energy powering the Sun, and by extension, the stars. The most obvious energy store for the Sun is also the immediate reason why it shines: it is hot. Could the Sun simply gently cool down over the immensely long timescale of the age of the Earth? The surprising answer is no: the Sun cannot cool!

The reason for this strange answer is something for which everyday observation does not prepare us. Hot objects familiar to us, such as a kettle, or a fireplace, or a car engine, can cool down to the temperature of their surroundings without significant structural changes. But the Sun is in a different position: it has to support itself against its own gravity, and its heat content is directly related to this fact – this is the reason it is hot.

We worked out that the Sun's central temperature is proportional to the ratio of its mass to its radius ($T_* \propto GM/R$). Now you can begin to see a problem emerging, for we would naively imagine that cooling down the Sun would involve a decrease of its central temperature. This in turn would mean that the ratio of mass to radius would decrease. There is no reason for the Sun's mass to change significantly as it cools, so to drop its central temperature, the Sun would actually have to *increase* its radius – all parts of the Sun would have to move slightly *outwards*. This is completely counter to our expectations: after all, it was only the fact that its centre was hot that stopped the Sun collapsing under its own weight, and yet here it is allegedly *expanding* as it cools down! There is clearly something wrong with this reasoning.

The error above is that we cannot consider heat energy in isolation from the other large energy store that the Sun has – gravity. Clearly, gravity *is* an energy source, since if it were not for the resistance of gas pressure, it would make all the Sun's gas move inwards at high speed. So heat and gravity are both potential sources of energy, and must be related by the need to keep the Sun in equilibrium. As the Sun tries to cool down, energy must be swapped between these two forms to keep the Sun in balance, changing the paradoxical deductions of the last paragraph. We reasoned that the centre of the Sun must be hot so that the gas particles there have energies that compete with gravity – that is, allow them to keep the Sun as an extended object, despite gravity trying to shrink it down indefinitely. But clearly (and fortunately), the heat energy inside the Sun is not enough to spread all of its contents out over space and destroy it as an identifiable object. The Sun is gravitationally *bound* – its heat energy is significant, but cannot supply enough energy to loosen gravity's grip, and unbind the Sun.

This means that when pressure balances gravity for any system (as in the Sun), the total heat energy T is always slightly less than that needed (V) to disperse it. In fact, it turns out to be exactly *half* of what would be needed for this dispersal, so that $2T + V = 0$, or $V = -2T$. The quantities T and V have opposite signs, because energy has to be supplied to overcome gravity, that is, you have to use T to try to cancel some of V. So T is positive and V is negative. There is nothing strange about a negative energy – the way to think of it is like a negative bank balance: you need to pay money into your bank to end up quits, and you need to supply energy to a star in order to overcome its gravity and disperse all of its gas to infinity. In line with this, the star's total energy (thermal plus gravitational) is $E = T + V = -T$, that is, the total energy is *minus* its thermal energy, and so is itself negative. That is, a star is a gravitationally bound object.

Whenever the system changes slowly enough that pressure always balances gravity, these two energies always have to be in this 1:2 ratio. So what happens if the Sun slowly shrinks by a small amount? Clearly, all the Sun's constituent atoms must be slightly closer together. This must mean that gravity is now a tighter force acting on the solar gas, and consequently that more energy would be needed to disperse the Sun. Since the total heat energy is exactly one-half of this increased quantity, it too must increase. This is not surprising: by shrinking, the Sun released some gravitational energy and put half of it into heat. But where did the other half go? We are assuming that gravity and heat are the two main stores of energy inside the Sun, and we have already accounted for their changes. So this excess energy must have left the Sun entirely. In other words, it must have been *radiated* by the Sun, as the Sun tried to cool. This reasoning shows that cooling, shrinking, and heating up all go together, that is, *as the Sun tries to cool down, its interior heats up*. As we asserted above, the Sun cannot cool. This is exactly what we see from the result that the star's total energy is just *minus* its thermal energy above. Because $E = -T$, when the star loses energy (by radiating), making its total energy E more negative, the thermal energy T gets more positive, that is, losing energy makes the star heat up.

This unusual behaviour results entirely from the fact the the Sun is a gravitationally bound system, that is, thermal pressure holds the Sun up against gravity, but there is not enough heat energy to disperse the Sun. The same thing happens in a simple system where a small mass orbits a larger one in a circular orbit (like a planet around the Sun). The equilibrium here is similar to that in the Sun. Gravity provides the centripetal pull keeping the planet in orbit, or colloquially, balances centrifugal force. So this simple system is another example of a gravitationally bound system. The analogue of the heat energy here is the kinetic energy of the orbiting mass – and once again, this turns out to be exactly one-half of the energy that would be needed for this mass just to escape the gravity of the larger mass and fly far away from it. If we take energy away from this system (i.e. try to 'cool' it) by moving the planet inwards to a closer orbit, gravity is stronger here, and the planet has to *speed up* to balance it, that is, it 'heats up'. Thus, once again, the attempt to cool the system actually leads to it heating up.

This result, that stars heat up when they try to cool, is central to understanding why stars evolve. In a sense, it is the basis of why we are here, for without stellar evolution there would be no life anywhere in the Universe. Unfortunately, this momentous result has a rather technical name; the *virial theorem*. The word 'virial' (coined by the 19th-century German physicist Rudolf Clausius) derives from the Latin word *vis*, meaning 'force' or 'energy'. The life of a star consists of trying to evade the dictates of this theorem. Its iron grip means that the interior of a star must at times inevitably shrink and heat up, rearranging the entire structure of the star. Its life can end only when the conditions of its interior change in such a way that the virial theorem no longer applies, and so no longer forces it to evolve.

Evidently, trying to use the Sun's internal heat to keep it shining would be a runaway process that only gets harder and harder: if the Sun were to slowly shrink and radiate, its interior would get hotter, losing more heat and reinforcing the process. To continue the analogy with a business losing money, it is as if paying out more money than you receive made you lose it even faster, rather like a gambler trying to compensate for a losing streak by laying ever larger bets. This kind of positive feedback obviously implies a time limit: the business and the gambler go bust faster than we feared, and the Sun changes to something unable to support life on Earth faster than we thought. We can estimate this time limit by dividing the Sun's total heat energy by the rate it loses energy into space. This kind of timescale appears so often in studying stars that it is called the *thermal timescale* or *Kelvin–Helmholtz timescale* of the star. Evidently, to work out this thermal timescale for the Sun, we need some estimate of its total store of heat energy.

Now you might think that this would require us to know all the details of the inside of the Sun, and even how these change as the Sun loses heat. This would clearly be a major undertaking. We would have to model all the complicated equations describing the physics of the Sun, and then write a program to solve them on a computer. Even for an experienced scientist, this would represent months of work. But we can sidestep these difficulties by realizing that we do not need to know the thermal timescale with great precision, but only well enough to compare it with estimates of the age of the Earth. So a very crude estimate of the thermal timescale may well be all we need: if this comes out far shorter than the estimated age of the Earth – a little over four billion years – then the idea that the Sun shines just by using its heat energy and slowly cooling down is plain wrong. We shall see that this is just what happens. To estimate the Sun's total heat energy, we adopt an almost childishly simple model of it: we pretend that all of it has about the same temperature – about 10^7 Kelvins – as its centre. Of course, this is not true, and very obviously wrong in the outer parts, where the temperature must be much closer to the surface value of around 6,000 K. But we can easily imagine that quite a large part of the Sun's gas mass is almost as hot as its centre, particularly as the gas is denser near the middle. So we are probably not making a huge error here.

The result of this extremely simple model is that the Sun's store of thermal energy is something like its total mass (2×10^{27} tons) multiplied by its central temperature (10^7 K) multiplied by a fixed number (the 'gas constant'). In this way, we estimate the Sun's thermal energy store as about about 2×10^{41} joules. (James Joule was a Manchester brewer who in 1845 experimentally measured how much heat energy is released from a hot object for each degree it cools. The brewery survived as an independent entity until 1974, and has recently reappeared.) To get the thermal timescale, we must divide this huge number by an estimate of the Sun's luminosity as it cools. We take this simply as the present luminosity of the Sun: after all, we want to know the time that this luminosity remains at this value. Dividing our estimate of the total heat energy by this luminosity now gives a thermal timescale of about 30 million years. In other words, the Sun's luminosity would begin to change noticeably on this timescale.

Now we can see that our extremely crude estimate, based on the simplest possible idealization of the Sun as a uniform ball, gives us a very clear answer. A timescale of 30 million years is about 150 times shorter than the 4.5 billion years we know the Earth has existed: the idea that the Sun is just a cooling hot object is wrong. It is *so* wrong that however bad our estimate of the Sun's thermal energy was, no error we might have made here, or in our luminosity estimate, can overthrow our conclusion.

Astronomers frequently make simple arguments of this type, and we shall see many examples in this book. They are effective because in astronomy we often consider situations that involve very disparate scales (of length, mass, time, and so on). Simple estimates like these are important, because they quickly show whether it is necessary or worthwhile to attempt an accurate calculation, with all the equations used properly. If we had constructed a very detailed computer model of the Sun, then despite all the months of work this would have involved, we would have reached the same answer. The discrepancy in thermal timescale and the age of the Earth might have come out slightly different, say a factor of 100 or 200 rather than our 150, but it would never alter our finding that the thermal timescale is by far the shorter of the two. So doing the big calculation would have been a massive waste of time and effort. The need to make simple estimates before attempting big computations is encapsulated in the semi-facetious phrase 'never start a calculation before you know the answer', usually attributed to the Princeton physicist John Wheeler.

A timescale of 30 million years is very long by human standards, but not nearly long enough. Even in the 19th century, this estimate was implausibly short as an age for the Earth, despite Kelvin's attempts to suggest otherwise. With modern estimates, the mismatch between the Sun's thermal timescale and the Earth's age is still worse. So nature must have a way out, and we need to find it. One or more of the assumptions we made in calculating the thermal timescale must have been wrong. This cannot be the connection between heat energy and gravitational energy, because abandoning this would mean that the Sun would collapse in an hour. Looking back, we can see that our hypothetical cooling Sun fell victim to the virial theorem simply because we assumed that it could only use heat to power its output. We would escape the age problem if the Sun had some other energy source which keeps it shining. This source has to be something like a hundred times larger than the solar thermal content.

In the 19th century, physics gave no hint of what could constitute the huge energy store that powered the stars, the largest coherent objects then known. Ironically, the answer turned out to involve a physical scale far smaller than the smallest – molecules and atoms – then contemplated. Three-quarters of a century would elapse before this discovery. We will discuss its momentous consequences in the next chapters.

Chapter 3
Life on main street

We started the last chapter by asking what supports the Sun against gravity, and ended by asking how it can shine for so long. Although gravity and electromagnetism correctly describe all the physics of stars on any length scale bigger than an atom, we know that these two forces are unable to provide the energy source that keeps the stars shining. So to find this source, we must look inwards, to physics on subatomic scales.

Atoms and nuclei

Atoms are the particles that make up all matter. The word 'atom' comes from ancient Greek, meaning 'cannot be cut'. The 'cutting' involved here means using forces familiar to us from everyday experience. There is not much mystery about what these forces are. Gravity is far too weak to be relevant on the scale of an atom, and all the other forces we meet in everyday life are electromagnetic in origin (this last statement will become clearer as we go on). So atoms are the smallest units of matter which interact only electromagnetically.

The structure of an atom is like a minuscule electromagnetic analogue of the Solar System. A positively charged nucleus has almost all the mass of the atom, and sits like the Sun at its centre.

Negatively charged bodies with much lower masses – electrons – orbit the nucleus like planets around the Sun. In an undisturbed atom, the electric charge of the nucleus and the total electron charge are equal and opposite, so that the atom has no net electric charge. The number of electrons in an uncharged atom like this (the atomic number) defines the chemical elements. Every undisturbed hydrogen atom in the universe has one electron, every helium atom two electrons, carbon six electrons, nitrogen seven, oxygen eight, and so on. The arrangement of the electron orbits is not arbitrary, but governed by definite physical rules we will discuss in detail later in the book. This orbital structure accounts for the line spectra of the elements. Each orbit corresponds to a precise and distinct energy for the electron in it. If light falls on the atom, some of the photons making up the light may have energies precisely equal to the energy gaps between these orbits. The electrons can absorb these photons, and lift themselves to more energetic orbits. This removal of photons of a certain fixed energy corresponds to a dark line in the spectrum. Eventually, the electron drops back from its higher-energy orbit, and re-emits the photon: if nothing else happens, this gives an emission line, as Kirchhoff and Bunsen found when they heated various elements.

Clearly, the occurrence of emission or absorption lines depends on what energy sources are available to excite the atom. In the outer layers of a cool star, there is very little radiation with the right energy to lift the electrons of hydrogen to higher orbits: there are too few photons in the radiation field with the right energy to do this. In a star of surface temperature about 10,000 K, there are plenty of these photons, so hydrogen lines are prominent. But if the star is significantly hotter, there are many photons capable of freeing the electron entirely from the clutches of the atom, and so transitions between the orbits giving hydrogen lines are rare. This explains the difficulty in working out the proportions of the various elements present in stars.

Atoms can bond together electromagnetically even though there is no net electric charge, simply because the electrons are not in precisely the same place as the nucleus, and orbit it rather than being static. Both effects lead to slight non-cancellations in the electromagnetic forces, and these allow atoms to stick together to make molecules. For example, an electron may end up orbiting both nuclei of a pair of atoms. These effects produce forces which are strong enough to hold atoms together, but become totally negligible at distances much greater than the size of a few atoms. This short-range nature of electromagnetic forces between atoms (and similarly, between molecules) is actually quite familiar to us. When you bang your knee against a hard object, you move in fractions of a millimetre from a situation in which there is no force between the atoms in your knee and those making up the object, to one in which the short-range electromagnetic forces between them become all too painfully evident.

We can already use our picture of the atom to make some interesting deductions. First, the whole of chemistry is simply the science of electromagnetic interaction of atoms with each other. Specifically, chemistry is what happens when electrons stick atoms together to make molecules. The electrons doing the sticking are the outer ones, those furthest from the nucleus. The physical rules governing the arrangement of electrons around the nucleus mean that atoms divide into families characterized by their outer electron configurations. Since the outer electrons specify the chemical properties of the elements, these families have similar chemistry. This is the origin of the periodic table of the elements. In this sense, chemistry is just a specialized branch of physics.

Second, atoms can combine, or react, in many different ways. A chemical reaction means that the electrons sticking atoms together are rearranging themselves. When this happens, electromagnetic energy may be released, because the new electronic configuration is tighter than the old one, or an energy supply may be needed, as the new configuration is looser. The energy here all ultimately appears as heat, so chemists call these types of reactions exothermic and endothermic respectively. Just as we measured gravitational binding energy as the amount of energy needed to disperse a body against the force of its own gravity, molecules have electromagnetic binding energies measured by the energies of the orbiting electrons holding them together. When we burn fuel in a fire, we are extracting electromagnetic energy by causing the electrons holding together the molecules in the fuel to change to more tightly bound configurations. The resulting ashes have lower fuel value, because it is more difficult to cause chemical changes to more tightly bound molecules that will make them more tightly bound still.

The same thing happens when we digest food: we extract electromagnetic energy from the food by changing the electron orbits in its molecules into more tightly bound orbits, and excrete the result as waste. We would rapidly run out of food if there were no process running in the opposite direction, transmuting this waste back into food. The inverse process is plant growth: plants absorb energy from sunlight to change tightly bound waste molecules into loosely bound food molecules. We will see that the process releasing energy inside the Sun is a vastly more energetic version of this movement from loosely bound to tightly bound structures, using a much stronger force than electromagnetism to do the binding.

The association of energy release with ever tighter binding suggests that we must look to even smaller lengthscales to find the Sun's energy source. Atoms are already small things: about 10^{-7}(one ten-millionth) of a millimetre, or equivalently 10^{-10} metres. But, as we have seen, changes of electronic binding only produce chemical energy yields, which are far too small to power stars. However, almost all of the tiny space of an atom is pure vacuum. Virtually all of the mass of an atom is contained in the nucleus, whose size is an incredible 10^{-12} (one trillionth) of a millimetre, or 10^{-15} metres. In terms of volume, the nucleus occupies only about one part in 10^{15} (one and fifteen zeroes) of the volume of the atom. Ernest Rutherford, the discoverer of the atomic nucleus, described this size disparity in the phrase 'the fly in the cathedral'. Even the nucleus is not the smallest thing we know of: it is itself made of smaller particles, protons and neutrons. These have similar masses, each almost two thousand times the electron mass, but each proton carries a positive charge equal and opposite to the electron, whereas the neutron has no charge, as its name suggests. Because of this property, and because it is very slightly more massive than the proton, one can think of a neutron as something like a proton and an electron combined together. The number of protons in the nucleus is exactly equal to the number of electrons in an uncharged atom of that element, and so also equal to the atomic number.

The simplest atom of all, corresponding to the lightest element, hydrogen, has a nucleus consisting of a single proton, balancing the charge of the single orbiting electron, and no neutrons. The nucleus of any other element has more than one proton – two for helium, and so on. But now we can see a potential problem. Protons all have positive charge, and so must repel each other. And this repulsion must be extremely strong because the nucleus is so small, and the force increases as the inverse square of the distance as we move inwards, just like gravity. How does the nucleus stay together despite this huge repulsion?

The answer is that there is another force of nature, the *strong nuclear force*, binding the protons (and indeed the neutrons) together, far more strongly than the electromagnetic repulsion pushes them apart. The force must have an extremely short range, acting only inside the nucleus, which is why it has no direct effect on everyday life. This finally is a candidate for the stronger interaction we have been looking for. We can imagine that some kinds of nuclei are more tightly bound together than others: if nature has a way of reorganizing the nuclear particles in these more tightly bound nuclei, energy will be released. Because the nuclear force is so strong, this *nuclear* energy yield is potentially far greater than the chemical energy release we get from transmuting molecules into more tightly bound configurations, which uses only the much weaker electromagnetic interaction. But note the momentous step we have taken here, at least in imagination, in the airy phrase 'reorganizing the nuclear particles into more tightly bound nuclei'. Reorganizing nuclei means altering the number of protons, and so transmuting one chemical element into another. When Rutherford's assistant, Frederick Soddy, first glimpsed this possibility, the reaction was instant: 'For Christ's sake, Soddy, don't call it transmutation. They'll have our heads off as alchemists.'

Nuclear fuel

Even accepting transmutation as a possibility, we still have two major steps to make before seeing it as realistic. First, we need to identify which nuclei are good candidates for energy release through nuclear transmutation. We cannot do this experimentally (as has happened with chemical reactions) because controlled nuclear transmutation is still a long way off, for reasons we will soon come to. So we need some way of estimating the binding energy of nuclei directly. Fortunately, there is a simple way to do this. It is a fundamental law of nature that mass and energy are just two forms of the same thing. This is the content of Einstein's famous relation $E = mc^2$. The equation comes from his theory of relativity, and tells us that removing energy E from something means that its mass is reduced, by the amount E/c^2, where c is the speed of light. Now remember the concept of binding energy. If something (say an atomic nucleus) is tightly bound, that means that you would have to supply more energy to disperse it than if it were loosely bound. Or equivalently, when you assemble the constituent protons and neutrons and let the strong nuclear force bind them tightly together, some energy gets lost – exactly the energy you would have to supply to disperse them again. Now by

Einstein's relation, this means that some *mass* was lost. So a more tightly bound nucleus has a lower mass than its constituents. The mass deficit is $m = E/c^2$, where E is the binding energy. It is fairly straightforward to measure experimentally the mass of a proton, a neutron, and all the atomic nuclei, and compare them. In every case but one, the nucleus has a lower mass than its constituent protons and neutrons, caused by the binding energy of the strong nuclear force. Only for hydrogen, which simply has one proton as its nucleus, is this mass deficit unmeasurably small: this is because no strong nuclear force is needed to bind this nucleus together. The next simplest nucleus is helium: this has two protons but also two neutrons. Its mass its about 0.7% less than the masses of its constituents.

Now the mass of the helium nucleus is suggestively close to four times the mass of a hydrogen atom. If the Sun and stars can find a way of making four hydrogen atoms combine to make a helium atom, we can see an amount of energy will be released equal to 0.7% of the total hydrogen mass times c^2. This may not sound a lot, but just consider the energy given by transmuting only *one kilogram* of hydrogen into helium. The speed of light is such a large number (3×10^8 metres per second) that one kilogram would release the staggering energy of 6×10^{14} joules. This would meet the energy consumption of the entire world for 8 minutes, or of the USA alone for about half an hour.

By comparison, burning one kilogram of oil produces about 4×10^7 joules, more than ten million times less. This is so small that the total mass deficit it produces is only about one part in almost two billion of the original mass. This is, of course, unmeasurably small, so the mass deficit method does not work for predicting which chemical fuels are most efficient. (And although you actually lose a little weight as you digest your food and use the energy, the amount is disappointingly small!) Converting hydrogen into helium is about 15 million times more effective than burning oil. This is because strong nuclear forces are so much more powerful than electromagnetic forces. (Sadly, this is why thermonuclear weapons are so much more destructive than conventional ones.) Converting hydrogen to helium is like an enormously more efficient way of burning fuel, so astronomers call it *hydrogen burning*. We will find later that other nuclear transmutations are important, and so we refer to the general process of converting loosely bound nuclei to more tightly bound ones as *nuclear burning*.

If hydrogen-burning powers the Sun, we can work out how much hydrogen fuel this process uses. We know that the Sun's output is a remarkably constant 3.8×10^{26} watts. Given the efficiency we worked out above, that each kilogram produces 6×10^{14} joules, and remembering that a watt is one joule per second, we find that the Sun must burn hydrogen at the rate

$$3.8 \times 10^{26} / 6 \times 10^{14} = 6.3 \times 10^{11} \text{ kilograms per second}$$

This sounds a lot, but remember that the Sun has a mass (mostly hydrogen) of 2×10^{30} kilograms. If it could burn all of its hydrogen, it could shine at its current rate for a time

$$2 \times 10^{30} / 6.3 \times 10^{11} = 3 \times 10^{18} \text{ seconds} = 10^{11} \text{ years},$$

since there are about 30 million (3×10^7) seconds in a year. This is definitely longer than the 4.5×10^9 years we know that the Earth has existed: the Sun need only have converted about 5% of its hydrogen to helium to have shone at its current rate for this time. Of course, the Sun's mass must have decreased as this happened, as it has lost energy. But since the lost mass is just 0.7% of the mass of the burned hydrogen, the total effect on the Sun is tiny: it must have lost 0.7% of 5% of its mass, or about one part in three thousand of its original mass.

All this suggests that hydrogen burning would answer the Sun's energy problem perfectly. But how do we know that this process really occurs in the Sun? What conditions are necessary to get four hydrogen nuclei (protons) to fuse together to make a helium nucleus? Remember that the strong nuclear force, which is where all the huge energies come from, operates only on tiny lengthscales of order the size of a nucleus. For the strong nuclear force to fuse the protons together requires the protons to get close enough for this force to work, that is, as close as the size of a nucleus, which we know is around 10^{-15} metres. If this happens, the strong nuclear force will grab the protons and combine them into a helium nucleus.

But we should also remember the reason we deduced the existence of this force: it is needed to overcome the electric repulsion force between the positively charged protons. It does this job just fine inside a nucleus, but has no effect outside it. So at distances only slightly bigger than a nucleus, the four protons will feel an enormous electric repulsion without any strong nuclear force to pull them together against it (see Figure 7). This electric force acts like a barrier trying to stop protons from getting close enough to combine. This is rather like having a golf hole precisely at the top of a mound (indeed some holes in so-called crazy golf are exactly like this). A ball has to climb to the top of the mound (i.e. overcome the electric barrier) before it can fall into the hole (be captured into a nucleus by the strong nuclear force). That means you have to hit the ball hard enough to reach the top, which is another way of saying that you have to give it enough energy to climb the mound. Here, because the electric repulsion grows like the inverse square of the distance, we know that it is the last bit of the climb that is the hardest, when the protons are separated by a distance similar to the size of the nucleus they are trying to make. The energy each proton needs is roughly this force multiplied by the distance it has to travel against it, that is, the nuclear size again. This energy is easy to work out, and is given by multiplying the repelling charges together and dividing by the nuclear distance. To initiate hydrogen burning, a star has somehow to supply this energy.

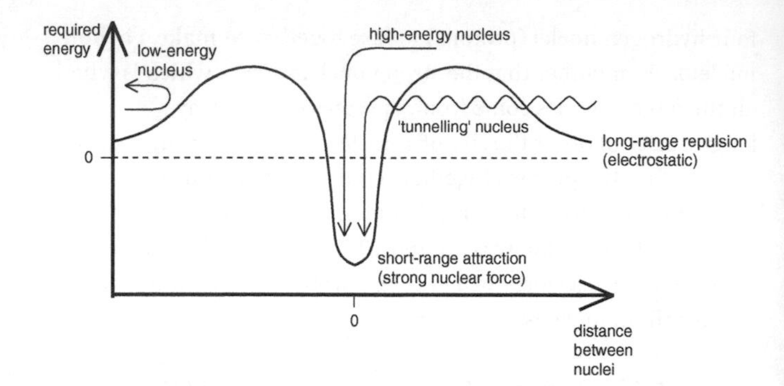

7. Nuclear fusion. A positively charged atomic nucleus repels other nuclei electrostatically (like charges repel) at large distances but attracts them strongly (by the strong nuclear force) if they can get close to it, causing fusion reactions. So an incoming nucleus must have enough energy to climb the rising, positive part of the potential energy curve to reach the steeply falling negative part near the nucleus. Nuclear fusion requires the nuclei to have high energies, which in practice means high temperatures

This process is exactly analogous to lighting a fire: we have to supply energy (for example, by striking a match) to make the fire go, because the chemical reaction which is the burning will not run unless the temperature is somewhat higher – high enough to make the relevant molecules break their current electronic bonds and start rearranging themselves in a more tightly bound configuration. Once this chemical reaction has started, it by itself produces enough heat to keep going. The need for an energy input makes fire manageable, of course: if the burning reaction needed no energy to start it, it would occur all the time. Just like fire, hydrogen burning needs an energy input to start it off.

Lighting the fire

But where can this energy come from? Inside a star, there is only one possible source. Remember, we worked out in the last chapter that the centre of a star like the Sun is a very hot place: the temperature there is around ten million degrees Kelvin. As we discussed, the temperature of a gas tells us the average energy of its constituent particles: at ten million degrees, this energy is thirty thousand times higher than at room temperature (about 300 degrees Kelvin). This is more than sixty times the energy needed to separate the electrons and protons in a hydrogen atom (this is called *ionizing* these atoms), so protons must be moving freely with this average energy. Is this enough for groups of four of them to overcome their electrical repulsion? Not even close: the energy needed for this would correspond roughly to an impressive hundred billion degrees Kelvin. This is discouraging, but since we have seen that nuclear burning would have all the right properties to power stars, we should ask if some variation of this idea might work. In particular, the temperature only specifies the *average* energy of the protons. Perhaps nuclear burning will work even if we have only a few protons with energies far above the average energy? The classical laws of thermal physics tell us that about one proton in a hundred has the required energy. But we need four protons like this, moving on converging paths, to make each helium nucleus, so this is a very rare event. In the last chapter, we saw that we have a reasonable idea of the density of hydrogen in the centre of a star like the Sun, so we can work out how often these rare events would occur, and estimate how much nuclear energy would be released. The result is far too small to explain the Sun's luminosity. So how does this work?

By the early years of the 20th century, astronomers had used the discovery of the nucleus to infer that hydrogen burning was an excellent fuel for powering the Sun and stars, but had no workable idea of how a star lights this nuclear fire. The breakthrough here came from a fundamental advance in physics, the move from the empirical 'old quantum theory' of the 1900s, to the modern quantum theory we use today, which emerged in the late 1920s. Among many other things, this revealed that an energy barrier of the kind we have encountered for hydrogen fusion is not absolute. In classical physics, which applies to everyday things, a body can only overcome an energy barrier if its energy exceeds the height of the barrier: so we have to hit golf balls hard enough to reach the height of the mound surrounding the crazy golf hole. In quantum mechanics, which applies on atomic scales or smaller, this is no longer so. A particle with lower energy has a small but finite probability of penetrating the barrier if it does so quickly enough. It is as if some of the golf balls tunnelled through the mound to reach the crazy golf hole. Indeed, this phenomenon is often called quantum tunnelling. Tunnelling like this is a direct consequence of the famous Uncertainty Principle. This is the conventional English translation of the original German name *Ungenauigkeitsprinzip*, which would have been far better translated as something like 'Inexactitude Principle', or 'Imprecision Principle'. (The conventional name has led far too many innumerate English-speaking writers to the erroneous view that physics is somehow fundamentally vague, or worse, simply a cultural concept.)

Whatever its name, the principle states in quantitative terms the degree of precision to which various pairs of quantities such as momentum and position, or energy and time, can – even in principle – be simultaneously measured, or indeed, be considered to have meaning. For an energy barrier such as we are considering, it says that a particle can penetrate the barrier with lower energy than the barrier height, provided that it is quick about it. The further below the barrier its energy is, the shorter the time available to penetrate it. The mathematical apparatus of quantum mechanics allows one to calculate precisely what fraction of a large number of bodies of a given energy will penetrate the barrier. Applied to our nuclear-burning problem, this means we can calculate quite accurately what the rate of hydrogen burning is for hydrogen gas of a given temperature and density. In fact, this calculation is very complex, because the simultaneous encounter of four hydrogen atoms is an extremely rare event. Hydrogen is turned into helium through a chain of reactions of pairs of particles, which are more common.

It turns out that there are two chains of reactions which can convert hydrogen to helium. The rate at which they occur is in both cases quite sensitive to the gas density, varying as its square, but extremely sensitive to the gas temperature. This latter result is a direct consequence of the importance of quantum tunnelling. If the temperature is below a certain threshold value, the total energy output from hydrogen burning is completely negligible. If the temperature rises only slightly above this threshold, the energy output becomes enormous. It becomes so enormous that the effect of all this energy hitting the gas in the star's centre is life-threatening to it. Remember that energy is related to mass. So being hit by energy is like being hit by mass: luminous energy exerts a pressure. For a luminosity above a certain limiting value related to the star's mass, the pressure will blow it apart. (This luminosity is called the 'Eddington limit', after its discoverer, Arthur Eddington: it appears as a footnote in his famous book on stellar structure, written in the 1920s.)

Hydrogen-burning stars

Clearly, a stable star like the Sun cannot have a luminosity above the Eddington limit. So the gas temperature in its centre cannot significantly exceed the threshold value. But equally obviously, the temperature cannot lie below the threshold, otherwise we would have no luminosity at all. The central temperature of the Sun, and stars like it, must be almost precisely at the threshold value. It is this temperature sensitivity which fixes the Sun's central temperature at the value of ten million degrees we deduced in the last chapter. All stars burning hydrogen in their centres must have temperatures close to this value. But in the last chapter, we deduced that this central temperature was roughly proportional to the ratio of mass to radius, i.e. $T_* \propto GM/R$. These two things are compatible only if *the radius of a hydrogen-burning star is approximately proportional to its mass*, or $R \propto M$. So a hydrogen-burning star with twice the mass of the Sun must have a radius which is also about twice as large, while one with half the mass will be roughly half as large, and so on.

You might wonder how the star 'knows' that its radius is supposed to have this value. This is simple: if the radius is too large, the star's central temperature is too low to produce any nuclear luminosity at all. By the reasoning of the last chapter, we know that the star will shrink in an attempt to provide the luminosity from its gravitational binding energy. But this shrinking is just what it needs to adjust the temperature in its centre to the right value to start hydrogen burning and produce exactly the right luminosity. Similarly, if the star's radius is slightly too small, its nuclear luminosity will grow very rapidly. This increases the radiation pressure, and forces the star to expand, again back to the right radius and so the right luminosity. These simple arguments show that the star's structure is self-adjusting, and therefore extremely stable, just as we argued that the Sun must have been for all of its lifetime. The basis of this stability is the sensitivity of the nuclear luminosity to temperature and so radius, which controls it like a thermostat.

It is remarkable that we can deduce such a powerful and far-reaching result as the relation between mass and radius from these ideas. But we can go further. We have seen that hydrogen burning needs rather special physical conditions, in particular a high temperature and reasonably high gas density. These conditions hold in the central region – the *core* – of the star. But the temperature and density must both decrease as we move away from the centre. So hydrogen burning only occurs in the core. The energy generated there has to make its way out through the envelope to the outer edge of the star. A star is certainly not transparent, so the nuclear luminosity released in the core has somehow to diffuse to the surface of the star, where it can finally escape into space. Any star is opaque to this radiation over quite short distances, which means that the radiation is constantly interacting with the gas of the star, being alternately emitted and absorbed by its atoms and electrons. Like any two hot things in close contact, the radiation and the gas must be in thermal equilibrium, that is, they must have the same temperature at each radius. The star's surface is much cooler than its centre, so this temperature steadily decreases as we move out through the star. If the decrease is gentle enough, the stellar material behaves like a blanket.

When you put blankets on your bed on a cold night, you are regulating the rate at which heat is conducted away from your body. Your aim is that this should just balance the rate at which you produce heat energy from food, so that you neither heat up nor cool down during the night. The most effective blankets are made of material which allows only slow heat loss, that is, material of low thermal conductivity. If you put a second blanket on your bed, the heat loss decreases. If this is now below the rate your body is producing heat, your body temperature will rise until the heat loss rate via conduction is again in balance with what you produce from food: more blankets make you hotter. (The actual rise in your body temperature is – fortunately – fairly small.)

Now a hydrogen-burning star is rather like this. The blanket is the envelope: all the gas outside the core, and its conductivity, is determined by the way radiation – light – interacts with its atoms. The star's nuclear luminosity is, of course, like the rate at which you use food to produce heat. However, there is an important difference between you and a star. As we have seen, a star's central temperature is *fixed* at about ten million degrees, and for a given stellar mass, completely determines its radius. The star is like someone lying in a bed with one blanket of fixed thickness and conductivity, but who nevertheless contrives to keep his body temperature constant. There is only one way out of this apparent paradox: unlike you, the star must adjust the rate at which it produces heat. That is, *there is a relation between stellar mass and luminosity*. If the stellar material is fairly hot, its conductivity is effectively constant, and one can show that the luminosity rises like the cube of the star's mass ($L \propto M^3$). For stars like this, twice the mass means a luminosity bigger by a factor eight. If the stellar material is cooler, its conductivity drops because its atoms retain more electrons which can interact with the stellar radiation field. This makes the luminosity increase even more rapidly with the star's mass, like the fourth or fifth power ($L \propto M^4$ or $L \propto M^5$). A star with twice the mass has a much higher luminosity in these cases: a factor sixteen or thirty-two, respectively.

Stars that behave like thermal conductors (blankets) in this way are said to be in *radiative equilibrium*. We can easily work out their surface properties, which are what we directly observe. For a given mass, we now know both the total luminosity and the radius of the star, and so its surface area. As we saw already in Chapter 1, physics tells us that the luminosity of an opaque hot body is given by multiplying its surface area by a constant times the fourth power of its surface temperature ($L = 4\pi R^2 \sigma T^4$, where σ is a constant number). The surface temperature T is of course far lower than its central temperature T_*. So if we know the luminosity of a star, we can now deduce its surface temperature. We recall from Chapter 1 that luminosity and temperature are the two axes of the fundamental Hertzsprung–Russell diagram (Figure 6) so the relation we find between these two defines some curve on this diagram. Doing the calculation, we find that the luminosity increases roughly as the seventh power of the temperature ($L \propto T^7$). This is the observed slope of the main sequence we met in Chapter 1, at least for stars that are not very cool. This result tells us that we are getting somewhere.

We worked out these relations for stars in radiative equilibrium, conducting radiation like blankets. But this is not true of all stars, because they are made of gas rather than being solid, and so can in principle move – *flow* – when heat is applied to them. We have to abandon the blanket analogy here, and instead consider heating water in a saucepan. Here we apply heat from below. At first, before the saucepan becomes hot, the water conducts heat gently upwards and does not move significantly. But once the base temperature is high enough, water starts to move bodily upwards in a chaotic way, ridding itself of some of its heat at the water surface, before sinking down again. A similar thing happens in stellar gas if the decrease of temperature with height exceeds a certain rate. Gas at lower depths becomes buoyant and rises bodily, giving up its heat at larger radii. This process is called *convection*, and occurs in the Earth's atmosphere too. If it occurs, convection is extremely efficient in transporting heat outwards in a star, and can carry a much higher luminosity than the radiative conduction process we considered before.

Convection is important in the cores of high-mass stars, because they have very high nuclear luminosities, which act rather like the hot base of the saucepan. It is also important in the outer regions (the envelopes) of low-mass stars. These have low luminosities and so low surface temperatures, so that we again have a situation in which the stellar temperature drops sharply as we move outwards. Convection persists until the stellar material becomes very opaque. This happens when the gas temperature drops to about 3,000 degrees Kelvin, for reasons connected with the atomic physics of these cool layers. Then the stellar luminosity can be conducted rather than convected because the temperature now decreases more slowly with radius. The end result is that all stars with outer convection zones tend to have very similar surface temperatures of around 3,000 degrees Kelvin. This agrees with the near-vertical slope of the main sequence for such cool stars.

Limits of the main sequence: degeneracy

We know now that nuclear burning of hydrogen to helium provides a very stable, long-lasting fuel in stars, and accounts for the observed features of the main sequence. This line on the Hertzsprung–Russell diagram (Figure 6) shows the position of stars burning hydrogen in their cores, the mass of the star increasing as we go up the diagram, like the luminosity. These stars – *main-sequence* stars, or *dwarfs* (not to be confused with white dwarfs) – have long lifetimes, and so are very numerous, making up most of the stars in the Universe. But there is one final question we should ask: what makes the difference between a star like the Sun, and a planet like the Jupiter? The answer lies in the nature of the pressure supporting these bodies against their own gravity. This changes fundamentally in matter that is very dense.

We have met the Uncertainty Principle earlier in this chapter. It says that the imprecision in the momentum of a particle and in its position cannot both be arbitrarily small. The degree of smallness specified here tells us that this principle is only important on scales smaller than the size of an atom. If matter is very dense, all its constituent particles are very close together and their positions are very tightly constrained. The Uncertainty Principle says that the momenta of these particles must have large imprecisions, which is only possible if these momenta can be large. Rapid motions like this mean that the matter exerts pressure, but one quite different from the thermal pressure we met in the last chapter, which came from the matter being hot. The new effect is called *degeneracy pressure*, and comes purely from the combination of very high density and the dictates of quantum theory. Degeneracy pressure appears in different species of particle at different densities; the heavier the particle, the higher the density. Since electrons are the lightest particles making up atoms, electron degeneracy pressure appears first as we go to higher densities.

If degeneracy pressure is what makes the difference between stars and planets, we have to look for conditions in which the density of gas in stars begins to become dense enough for it to appear. Density (ρ) is just mass divided by volume, and the volume of a sphere like a star is proportional to the cube of the radius ($\rho \propto M/R^3$). For a low-mass main-sequence star, we already know that the radius is proportional to the mass, so the mean density of this kind of star is proportional to mass divided by mass cubed, or equivalently, inversely proportional to the square of the mass ($\rho \propto M/M^3 = 1/M^2$). A main-sequence star with half the mass of the Sun is on average four times denser, and one with one-tenth of its mass is one hundred times denser. At about this mass, electron degeneracy pressure starts to become significant in holding up the star against its own gravity. In other words, these stars do not need to raise their central temperatures to support themselves. But this need was precisely the effect that led to them initiating hydrogen burning. So for masses less than about 0.1 times the Sun's mass, a ball of hydrogen does not heat up and glow – it is not what we think of as a star at all, but something much more like a planet. In fact, planets are less massive still (about one-hundredth of the Sun's mass), and these more massive objects just at the bottom of the main sequence are called 'brown dwarfs'.

So the lowest possible mass for a main-sequence star is about one-tenth of the Sun's. What about the highest mass? Very massive stars must be very luminous, since we saw that luminosity grows strongly with mass (as $L \propto M^3$) for hot stars. For masses about sixty times the Sun's, this luminosity is about equal to the Eddington limit we met a few pages earlier. This tells us that radiation pressure must now be important, and change the star's structure. The extra pressure prevents the central density of the star increasing so rapidly with total stellar mass, and so slows the increase of luminosity with mass. These stars arrange themselves to have luminosities close to their Eddington limits, so luminosity now increases proportionally to mass ($L \propto M$) rather than its cube. There is no reason in principle why such stars cannot exist with quite high masses, and several are known at more than one hundred solar masses. But in practice, such stars are rare, probably because it is not easy to collect a gas mass of this order and form it into a star. In any case (as we shall see in the next chapter), the high luminosities of these stars mean that they do not live very long, and so are rare for this reason too.

Compared with these limits of the main sequence, the Sun is indeed a very average star: its mass is neither exceptionally large nor unusually small.

Chapter 4
Cooking the elements

Growth and change

We now know that most stars shine by nuclear burning of hydrogen to helium. As this process consumes the hydrogen in their centres, the stars must *evolve*. Can we observe this process? At first sight, this question might seem ridiculous – human life is far shorter than the age of the Sun, and presumably of the other stars too, so how can we learn anything of their development over time? But there is a suggestive analogy here. Imagine taking a walk through a forest. You see a variety of trees, small, tall, some in full leaf, and others bare. If no-one had ever told you how trees grow, you might at first wonder if there was any relation between these different trees. Eventually, you might guess that the small sappy trees were young, the tall ones mature, and those without leaves, or broken, or fallen, were somehow the endpoints of a tree's life. And you might infer all this without waiting to see any individual tree change from one stage to another. This analogy, first given by Eddington, shows that we should be able to deduce the way stars evolve by looking at large numbers of them, and comparing our results with what physics predicts for this evolution.

Evolution is about time, so the first thing we need is an accurate estimate of how long a star can shine by burning hydrogen in its core, its *nuclear lifetime*. In the last chapter, we made a first attempt at this by asking how long the Sun would need to burn *all* of its hydrogen at the rate producing its current energy output. But it may not be able to do this, because the act of burning must alter the state of the core by piling up helium 'ashes' there. So the nuclear lifetime might be shorter than our first estimate.

Hydrogen burning produces a dense and growing ball of helium at the star's centre. Once again, the star has a weight problem to solve – the helium ball feels its own weight, and that of all the rest of the star as well. A similar effect led to the ignition of hydrogen in the first place – the central hydrogen became so dense and hot that enough hydrogen nuclei came close enough to overcome their electrostatic repulsion and fuse together under the strong nuclear force. Could a similar thing work for helium? We will see that it does in the end. But not just yet – helium nuclei have two protons (and two neutrons), and so twice the electric charge. Their electrostatic repulsion is $2 \times 2 = 4$ times stronger than for hydrogen, requiring a higher temperature to overcome it and set off fusion reactions between helium atoms. The helium ashes of hydrogen burning are too cool for this, so there are no nuclear reactions in the helium core at this stage. This means there is no way to keep any part of the core hotter than its surroundings, and it must rapidly adopt the temperature of the surrounding envelope all the way through its structure.

This 'dead' helium core resembles a small star of growing mass at the centre of the main one. To resist the weight of the envelope, the core pressure must match that of the hydrogen envelope just outside it. The core's central pressure must be higher than this, as it has to cope with its own weight as well. We saw that gas pressure is proportional to the number of gas particles in a given volume (the *number density*) multiplied by the temperature. But the helium core has the same temperature as the envelope just outside. So to have higher pressure in the centre of the core, the helium number density must be higher than in the hydrogen envelope. The central number density is proportional to the core mass divided by its volume, which goes as the cube of its radius. And the constant temperature in the core means that this radius is directly proportional to the core mass ($R_{core} \propto M_{core}$) – remember our argument that the central temperature of a star goes as $T_* \propto M/R$, and a constant central temperature (there fixed by hydrogen burning) required $R \propto M$. So now we can see what happens as the core mass grows. Let's imagine that the core mass has doubled. Then the core radius also doubles, and its volume grows by a factor $2 \times 2 \times 2 = 8$. This is a bigger factor than the mass growth, so the density is $2/(2 \times 2 \times 2) = 1/4$ of its original value. We end with the surprising result that as the helium core mass grows in time, its central number density *drops*. (Mathematically, this is just density $\propto M_{core} / R^3_{core} \propto 1 / M^2_{core}$.) Because pressure is proportional to density, the central pressure of the core drops also ($P \propto T_*/M^2$).

Since the density of the hydrogen envelope does not change over time, we see that the helium core becomes less and less able to cope with its weight problem as its mass increases. The problem is made worse by the fact that the core is made of helium, so that the same number density of helium corresponds to a higher mass than for hydrogen. The end result is that once the helium core contains more than about 10% of the star's mass, its pressure is too low to support the weight of the star, and things have to change drastically.

We will work out shortly what happens to the star. But we can already work out its nuclear lifetime. Using the Sun as an example, we take just 10% of our first crude estimate in Chapter 3 of about 10^{11} years, because only about 10% of the hydrogen gets burned, and find a *main-sequence lifetime* of about 10^{10} (ten billion) years (or $t_{MS} = 10^{10}$ years). As we shall see, the main sequence phase is by far the longest that a star experiences. So the Sun is just about half way through its total lifetime now.

We can extend this estimate to main-sequence stars of different masses. Again, we take the hydrogen fuel store to be just 10% of the total stellar mass. The star burns this at a rate given by its main-sequence luminosity. In the last chapter, we found that the more massive the star, the higher the luminosity. For stars more massive than the Sun, the luminosity goes like the cube of the mass ($L \propto M^3$). Scaling everything to the Sun, we see that massive stars have much shorter main-sequence lifetimes, decreasing like the inverse square of their masses ($t_{MS} \propto 1/M^2$). A ten solar-mass star has only one-hundredth ($1/(10 \times 10)$) of the Sun's lifetime, that is, about 10^8 (one hundred million) years. One the other hand, stars less massive than the Sun have feebler luminosities and so much longer lifetimes. A star near the minimum main-sequence mass of one-tenth of the Sun's has an unimaginably long lifetime of almost 10^{13} years, nearly a thousand times the Sun's. All low-mass stars are still in the first flush of youth. This is the fundamental fact of stellar life: massive stars have short lives, and low-mass stars live almost forever – certainly far longer than the current age of the Universe.

The nuclear timescale is the longest of three fundamental timescales which together govern the entire life story of a star. The next shortest is the thermal, or Kelvin–Helmholtz, timescale, which measures how long the star can go on shining without any nuclear reactions to resupply the lost energy. The shortest is the dynamical timescale, the time for the star to collapse in on itself if it had no pressure support. We have met all three of these timescales for the Sun. The nuclear time is ten billion years, the thermal timescale is thirty million years, and the dynamical one (see Chapter 2) just half an hour. The striking fact here is how very different these times are. The nuclear time is 300 times the thermal time, which is itself almost 10^{12} dynamical times. This clear hierarchy holds for all stars, and makes their evolution easy to follow.

Each timescale says how long the star takes to react to changes of the given type. The dynamical time tells us that if we mess up the hydrostatic balance between pressure and weight, the star will react by moving its mass around for a few dynamical times (in the Sun's case, a few hours) and then settle down to a new state in which pressure and weight are in balance. And because this time is so short compared with the thermal time, the stellar material will not have lost or gained any significant amount of heat, but simply carried this around with the same bits of mass which had them before the star was disturbed. But although the star quickly finds a new *hydrostatic* equilibrium, this will not correspond to *thermal* equilibrium, where heat moves smoothly outwards through the star at precisely the rate determined by the nuclear reactions deep in the centre. Instead, some bits of the star will be too cool to pass all this heat on outwards, and some will be too hot to absorb much of it. Over a thermal timescale (a few tens of millions of years in the Sun), the cool parts will absorb the extra heat they need from the stellar radiation field, and the hot parts rid themselves of the excess they have, until we again reach a new state of thermal equilibrium. Finally, the nuclear timescale tells us the time over which the star synthesizes new chemical elements, radiating the released energy into space.

The hierarchy of timescales means that stars are in various states of both hydrostatic and thermal equilibrium for almost all of their lives, with very rapid changes between these different states. Here 'very rapid' means 'very rapid compared with the nuclear lifetime' – we shall see that stars like the Sun actually spend several tens of millions of years out of thermal balance, but this is tiny compared with the ten billion years of thermal equilibrium preceding it, and the hundreds of millions of years that follow. So unless the non-equilibrium states are unusually bright, we should not expect to see many stars in these states. In terms of our analogy with the forest, we do not expect to see many trees in the act of falling, but only ones either standing or fallen. The only exceptions occur when some effect continually disturbs the equilibrium the star is trying to reach. For example, we will see that at certain (short) stages, stars are likely to pulsate – regularly expand and contract – because this motion itself causes changes that feed energy into the oscillations and keep them going. Other cases occur when the star is continually disturbed by a close companion star which adds or removes gas. In the forest analogy, these are rather like finding a dead tree whose fall to the ground has been arrested by the branches of nearby trees.

Giants

So what happens when a star can no longer stay on the main sequence? Once the helium core reaches about 10% of the star's mass, we have seen that it can no longer support the weight above it. So it must contract and become much denser. If the star's mass is less than about twice the Sun's, the core is dense enough that the degeneracy pressure force we discussed in Chapter 3 is able to slow the collapse. In more massive stars, the core shrinks to the point when the centre gets hot enough for gas pressure to do this. But heat leaks out, just because the core centre is hotter than its edge, so a slow shrinkage continues in both cases. During this slow squashing, the core and the envelope are always close to hydrostatic balance, so we know from Chapter 2 that the gravitational and thermal energies of the star must stay in the 2:1 ratio specified by the virial theorem ($2T + V = 0$). And provided that the star is close to thermal equilibrium while this squashing happens, the total stellar energy (gravitational plus thermal, $E = T + V$) stays fixed. These two results mean that the gravitational (V) and thermal (T) energies of the whole star are each separately unchanged during the process.

But squashing the core increases its (negative) gravitational energy: the star has a lot of mass close together in the core, making the gravitational attraction very strong (i.e. V gets more negative). To conserve the star's total gravitational energy, its envelope must *lose* gravitational energy – which means it must expand. But gravity is strong in the core and weak in the envelope. If we think of energy as money, this means that contracting the core pays out a lot, but expanding the envelope is very cheap. To keep things in balance – to conserve gravitational energy – the star has to expand a great deal. So the star must get much larger in overall size. At the same time, the core heats up as it contracts, forcing the envelope to cool down to conserve the total thermal energy. These simple arguments tell us that after a star burns about 10% of its hydrogen to helium, it must expand to become a very large and cool object. A star like the Sun will eventually expand to about two hundred times its present size, just about reaching the Earth's orbit.

At the same time, a shell of hydrogen just outside the core gets hot enough to start burning itself, and this makes the star more luminous than when it was on the main sequence. But remember that the total radiation – the luminosity – of a surface is proportional to the total area times the temperature multiplied by itself four times ($L = 4\pi R^2 \sigma T^4$). And here, the surface area radiating this luminosity is now so large that the temperature of the visible surface becomes quite low, and the star gets very red. This is a *red giant*.

We met red giants in Chapter 1. They occupy a near-vertical line on the extreme right (low-temperature) side of the HR diagram (Figure 6). This line meets the main sequence line, occupied by the dwarfs, at a point where the stars have masses similar to the Sun, but giants are well separated from dwarfs for larger masses. The empty region between these two families is the *Hertzsprung gap*. But from our discussion above, we know that this region should in fact be occupied by stars that are growing their radii from main sequence values to the far larger red giant sizes. The reason we do not see these stars goes back to our discussion of 'rapid' changes between long-lived equilibrium phases a few paragraphs above. Remember, the huge radius growth is the direct result of strong thermal (and gravitational) energy imbalance – these stars are far from thermal equilibrium. Putting this right, and restoring thermal balance, takes a thermal timescale. So these hidden stars cross the Hertzsprung gap in about a thermal timescale. This is far shorter than their time on the main sequence, and they are no brighter, so we expect to see far fewer of them. In terms of the stellar 'forest', we do not see many dead trees in the act of falling. The red giant phase of a star's life is significantly shorter than on the main sequence, but a star of a given mass is much brighter when it is a giant than when it was a dwarf. As a result, we can see giants out to much larger distances, where the dwarfs are invisible, increasing the giant fraction among the stars we do see. A large fraction of the naked-eye stars in the night sky are actually red giants, Betelgeuse being the most prominent example.

The red giant phase is not the end of a star's life story. Deep in the core, the density and temperature slowly rise until helium is able to burn, making carbon and oxygen. This heats up the helium core and makes it expand. The processes that made the envelope swell when the core contracted at the end of hydrogen burning now go into reverse: to conserve gravitational and thermal energy, the envelope now *shrinks*. The star now appears smaller and hotter, and so moves to the right in the HR diagram (Figure 8). This phase of core helium burning is considerably shorter than the hydrogen equivalent (the main sequence) because the nuclear binding energy release in transmuting helium into carbon and oxygen is less, and the stellar luminosity is greater, than in that phase. It is in this stage of core helium burning that stars develop a tendency to pulsate, as we mentioned earlier. Although this is not important for the star's evolution, this property turns out to be extremely useful in measuring cosmic distances, as we shall see in Chapter 7.

So far, our original star has added to its helium content and produced carbon and oxygen. In doing this, it left the main sequence, and swung right and then left in the HR diagram. The next stages are easy to imagine in outline but complex to describe in detail. Eventually, both carbon and oxygen are ignited, producing magnesium, sodium, neon, and silicon. Silicon then forms the basis of still more complex transmutations. But building

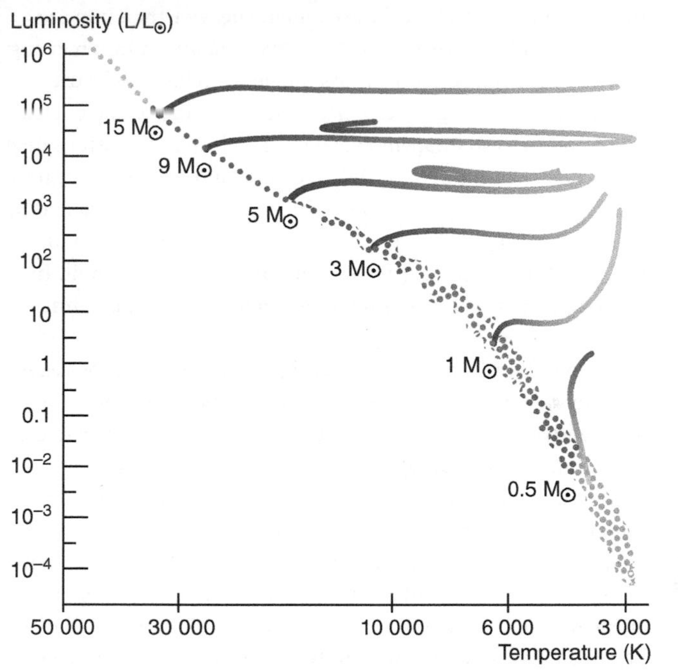

8. Evolutionary tracks on the Hertzsprung–Russell diagram. Main-sequence stars move to the right and become red giants after ending core hydrogen burning. The speed of this motion depends on the star's mass: slow for low-mass stars, and high for massive stars. Lower-mass stars move up the red giant branch as their degenerate helium cores grow, and eventually evolve from red giants into white dwarfs (lower left-hand corner, track not shown). Massive stars move back and forth across the diagram before ending their lives as neutron stars or black holes

up heavier elements by fusing lighter ones has a natural limit. Until now, this fusion has always given out some nuclear energy, because the strong nuclear force binds the new nuclei together more tightly than the old ones. But the iron nucleus is the most tightly bound nucleus of all. So once iron is reached, further fusion would require an energy *input* rather than releasing energy. So fusion no longer works. All the elements heavier than iron are made by a different process. Carbon, oxygen, and silicon burning produce a lot of free neutrons. As the neutron has no electric charge, it is easily captured to heavy nuclei by the strong nuclear force, and this can happen several times. The extra neutrons may not be stable, and can decay into a proton and an electron (plus an anti-neutrino, a neutral particle which interacts only through the weak nuclear force). The electron escapes and leaves a proton bound to the nucleus, raising its nuclear charge by one unit. In this way, stars are able to produce elements heavier than iron.

As we have by now come to expect, igniting these reactions causes structural changes in the stars involved, and makes them alternately expand and contract, swinging back and forth across the HR diagram. If the star's mass is large, all the helium is eventually exhausted in the core. Then this contracts still further, forcing an even larger expansion of the envelope. These huge stars are called *supergiants*. Their cores are layered like a giant onion: each layer has a different chemical composition, with lighter elements lying above heavier ones. Outside this core is a relatively homogenous envelope, still mainly hydrogen, but now enriched (or polluted) with many of the heavier elements produced within. This has happened because convection – the boiling motions that can transport energy outwards in a star – has stirred the various onion layers, and particularly the outer envelope, whenever this gets very large and cool. We will see in the next chapter how these elements are spread through space, producing all the phenomena familiar to us, including our own bodies.

The vast sweep of this chain of discovery is worth some thought. Starting from a simple ball of hydrogen and helium, the laws of physics inexorably drive the creation of all the other chemical elements. At each stage, the iron dictates of the virial theorem – just gravity and thermodynamics – force the star to look for new energy sources. Each one makes new elements. So the stars shine, and make the stuff that makes us.

Chapter 5
Stellar corpses

Endings

Stars live a long time, but must eventually die. Their stores of nuclear energy are finite, so they cannot shine forever. But how do stars end their lives? We have seen that they are forced onwards through a succession of evolutionary states because the virial theorem connects gravity with thermodynamics and prevents them from cooling down. So main-sequence dwarfs inexorably become red giants, and then supergiants. What breaks this chain? Its crucial link is that the pressure supporting a star depends on how hot it is. This link would snap if the star was instead held up by a pressure which did not care about its heat content. Finally freed from the demand to stay hot to support itself, a star like this would slowly cool down and die. This would be an endpoint for stellar evolution.

We have already met this kind of pressure in Chapter 3. Electron degeneracy pressure does not depend on temperature, only density. If the matter density is high, electrons are close to each other, and the Uncertainty Principle forces them to move rapidly, exerting a strong pressure which has nothing to do with temperature. So one possible endpoint of stellar evolution arises when a star is so compressed that electron degeneracy is its main form of pressure. We have already seen that the helium cores of low-mass stars become degenerate as the stars leave the main sequence and evolve into red giants. These cores later expand when helium starts to burn, but get squashed and strongly degenerate again once the core has become a mixture of carbon, neon, and oxygen. By this point, the star is a supergiant, and a lot of its mass is in a hugely extended envelope, several hundred times the Sun's radius. Because of this vast size, the gravity tying the envelope to the core is very weak. The inverse square law tells us that it is ten to a hundred thousand times weaker than at the Sun's surface – the envelope has almost no *weight*. Even quite small outward forces can easily overcome this feeble pull and liberate mass from the envelope, so a lot of the star's mass is blown out into space. Eventually, almost the entire remaining envelope is ejected as a roughly spherical cloud of gas. The core quickly exhausts the thin shell of nuclear-burning material on its surface. Now gravity makes the core contract in on itself and become denser, increasing the electron degeneracy pressure further. The core ends as an extremely compact star, with a radius similar to the Earth's, but a mass similar to the Sun, supported by this pressure. This is a *white dwarf*. Even though its surface is at least initially hot, its small surface means that it is faint. These stars appear in the bottom left-hand corner of the HR diagram (Figures 6 and 8).

With so much mass in such a small space, a white dwarf has a staggeringly high density: a teaspoonful of white dwarf weighs about a ton. A white dwarf is similar to the Earth in size, so its much larger mass makes the gravity at its surface more than a hundred thousand times stronger. White dwarfs cannot start nuclear reactions, so eventually they must cool down and become dark, cold, dead objects. But before this happens, they still glow from the heat energy left over from their earlier evolution, slowly getting fainter. Astronomers observe many white dwarfs in the sky, suggesting that this is how a large fraction of all stars end their lives. Their masses cover a range reflecting their past evolution, heavier main-sequence stars leaving more massive white dwarf remnants.

Neutron stars and black holes

But not all stars can end as white dwarfs. The degeneracy pressure holding these stars up against their intense gravity requires high densities, so a white dwarf of higher total mass is actually denser, and smaller in radius, than a lower-mass one. Its electrons are forced closer together, so by the Uncertainty Principle they have to move faster. In a white dwarf with a mass only slightly larger than the Sun's, the electrons have to move at almost the speed of light. Then something important happens: the electrons' kinetic energy is similar to their mass energy, specified by Einstein's famous relation $E = mc^2$.

We used this relation in earlier chapters to work out how much energy was released in transforming one type of atomic nucleus into another – that is, $E = mc^2$ told us that mass gave us energy. But the relation also works the other way round: energy gives us mass! And since mass has weight, this means that *energy has weight*. In other words, once the electrons are moving at almost the speed of light, increasing their density and pressure no longer helps to hold up the star against gravity – it just adds to the weight that the pressure must support.

The result is inevitable: once a white dwarf is so massive that its electrons reach speeds close to light, electron degeneracy pressure cannot support it. This fixes a maximum possible mass for any white dwarf. Physical laws allow us to calculate this mass. It turns out to have an extremely interesting value – only about 1.4 times the mass of the Sun. A young Indian astrophysicist, Subrahmanyan Chandrasekhar, was the first to derive this result clearly in this form in the course of a 1930 sea voyage to England, where he was to study, although slightly cruder versions were given earlier by both Anderson and Stoner. The full significance of this result was not realized until much later. Chandrasekhar received the Nobel Prize for this discovery in 1983, and the critical mass is named after him.

The existence of the Chandrasekhar mass solves one problem but creates another. Its value is large enough for us to be confident that stars with low masses can end their evolution as white dwarfs. Indeed, our discussion of stellar evolution showed that low-mass stars in a sense grow these strange stars in their interiors for a long time. But astronomers know of many stars with much higher masses than Chandrasekhar's critical value. These stars cannot end up as white dwarfs unless they lose most of their mass. We have seen that stars do lose a lot of mass as they evolve, but can they all lose enough to end with a mass below the Chandrasekhar value of 1.4 solar masses? We will see later how observations give us an answer to this question. It is a clear 'no'. Stars with an initial mass more than about seven times the Sun's cannot end as white dwarfs.

Stars more massive than this must end differently. Near the end of their time as supergiants, nuclear reactions have turned the central parts of their cores into pure iron. This is the most tightly bound atomic nucleus, so there is no hope of any further reaction supplying the heat necessary to stave off collapse. When this collapse finally begins, the core is already above the Chandrasekhar limiting mass, and electron degeneracy pressure is powerless to stop the infall. Matter is squeezed inexorably tighter by gravity, falling almost freely inwards. This is a dramatic moment in the life of the star. For almost all of its previous existence, changes happened extremely slowly – the star spent hundreds of millions of years burning hydrogen on the main sequence, and millions of years evolving through the later stages. Yet here the star's structure changes drastically on the time it takes matter to fall inwards, which in the deep interior is only a few seconds or less.

It is obvious that this infall process must release huge amounts of energy. Where does this go? For once, this question is simple – we know that normal gas pressure, caused by matter moving about chaotically, is far smaller than electron degeneracy pressure. So there is nothing to stop the energy simply going into the motion of the gas particles. This is another way of saying that the temperature rises almost without limit. This thermal energy becomes comparable with the binding energy of the iron nuclei, and these disintegrate into helium nuclei and free neutrons. The helium nuclei themselves break into neutrons and free protons. All the time, the collapse is squashing all these particles ever closer together. Eventually, the density becomes high enough for the free protons to capture free electrons and turn into neutrons. This process removes thermal energy and reduces the total number of particles, so the pressure drops still more and the collapse continues, squashing this neutron gas. Eventually, these neutrons are pushed so close together that they become degenerate just like the electrons in a white dwarf. The Uncertainty Principle forces the neutrons to move rapidly, and this neutron degeneracy pressure can potentially stabilize the core, just as electron degeneracy pressure does for helium cores, and for white dwarfs.

But there is a significant difference. The mass of a neutron is almost 2,000 times that of an electron, so for a given momentum, the Uncertainty Principle requires a neutron to move about 2,000 times slower than an electron. The pressure the gas exerts is roughly proportional to the number density of the particles multiplied by the rate at which they transfer momentum. So for a given density, our sluggish degenerate neutrons give *lower* pressure than electrons. Then to exert enough pressure to balance a certain weight, a degenerate neutron gas must be *much* denser than an electron gas. A degenerate neutron core with a mass of about 1.5 times the Sun stabilizes at a radius of about 10 to 20 kilometres, far smaller than a white dwarf.

This tiny size means that gravity at the core boundary is far stronger than at the surface of the star, by the inverse square law. It is a staggering 5 billion times stronger than at the surface of the Sun. Since the core collapsed from a size considerably larger than its current 10–20 kilometres, it must have released a huge amount of gravitational energy. For a core mass about 1.5 solar masses, this is an almost unimaginable 3×10^{46} joules. To get some idea of what this number means, we compare it with the Sun's total radiation output as it burns hydrogen on the main sequence. The Sun's luminosity is about 4×10^{26} watts, and its main-sequence lifetime is about 10^{10} years, or about 3×10^{17} seconds. Since one watt is a joule per second, we find that the Sun's total output over its 10^{10}-year main-sequence lifetime is about 10^{44} joules. So the energy released by the collapsing core of a massive star is about 300 times greater! To make things still more spectacular, most of this energy is released in the time the core takes to collapse to its final radius from one only twice as large. The core collapses at almost the free-fall speed, which is about 10^5 kilometres per second (about one-third of the speed of light), so the timescale of the gravitational energy release is only 10^{-4} seconds (i.e. one ten-thousandth of a second).

The obvious question is where this huge energy goes. To answer this, we remember that the net effect of all the complex nuclear reactions has been to transform all the protons in the original iron core into neutrons. For a core mass of about the Sun's mass, this means that roughly 10^{57} protons and 10^{57} electrons have to be turned into the same number of neutrons. Each time this happens, the reaction emits a very weakly interacting, almost massless, particle – a *neutrino*. (This process is the inverse of the beta-decay of a free neutron into a proton, electron, and anti-neutrino.) Since they interact so weakly, neutrinos can easily pass through huge amounts of ordinary matter without being slowed or deflected. So almost all of the 10^{57} neutrinos escape the star completely, and this is where the vast bulk of the gravitational energy release goes.

If this was all that happened, the cataclysmic events we have been describing would leave surprisingly little trace. The weakness of neutrino interactions with matter make them extraordinarily hard to detect. The Sun's nuclear reactions also produce significant numbers of neutrinos, and the flux at Earth is about 7×10^{10} per second on each square centimetre. This means that about 4×10^{14} solar neutrinos are passing through your body every second. The flux of primordial neutrinos left over from the Big Bang is even about forty times more than this. Yet only the most heroic efforts, involving huge detectors (typically a giant tank of dry-cleaning fluid) placed underground in disused mines, allowed the detection of solar neutrinos, winning Ray Davies a Nobel Prize in 2002. So neutrino interaction with matter really is weak.

But weak as this connection is, when neutrinos have to pass through the entire envelope of a massive star, some of them do interact with it. A few per cent of the total neutrino energy ends up in the envelope, giving this about 10^{45} joules. This is a few times more than the binding energy, that is, the energy needed to lift the envelope into space against the core's gravity. The entire envelope is blown out into space with velocities of tens of thousands of kilometres per second. This gas collides with the interstellar medium and drives powerful shock waves which heat both the ejected and swept-up interstellar gas. This gas now radiates at the huge rate of about 3×10^{10} times the Sun's luminosity for about a year, after which it slowly decays as the ejecta sweep up more gas and decelerate. This is a core-collapse *supernova*.

Needless to say, supernovae are very easy to see at enormous distances, and astronomers now detect several hundred per year. All of these are quite distant from us, the nearest being one in 1987 in the Large Magellanic Cloud, a small galaxy quite close to the Milky Way. Supernovae like this in nearby galaxies can be seen with the naked eye. The Crab Nebula (Figure 9) is the remnant of a supernova explosion in the Milky Way, and appears in Chinese records as occurring on 4 July 1054. But although supernovae are very bright in optical (or more generally, electromagnetic) radiation, we should remember that this is completely dwarfed by

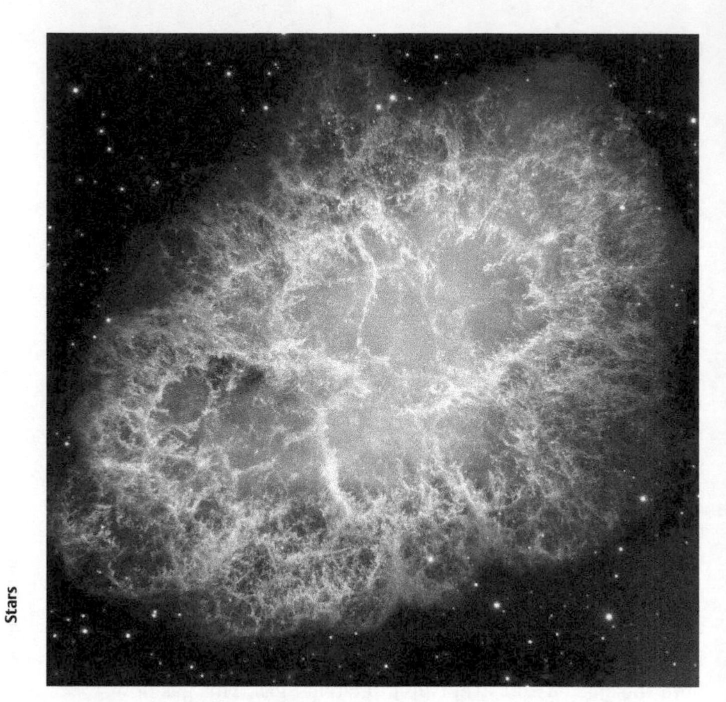

9. The Crab Nebula. The nearest supernova remnant to us, this is the debris of a star that exploded on 4 July 1054 AD. The core of the star has become a neutron star, which is detectable as a pulsar near the centre of the nebula

the neutrino emission, which is a hundred times greater in total energy. But the weakness of neutrino interactions makes it a real challenge to detect this emission from a supernova. Masatoshi Koshiba also shared in the 2002 Nobel Prize for his role in leading the first such detection, of the 1987 supernova in the Large Magellanic Cloud.

Supernovae have far greater significance than their spectacular appearance. In expelling the entire envelope of massive stars, they enrich the interstellar gas with elements heavier than the primordial hydrogen and helium from the Big Bang. This is the only way of making these elements, and without supernovae they would remain buried within the stars that made them. Yet spectrographs show us that these elements are present wherever we look in the Universe. And of course, we know that much of the Earth, and indeed our own bodies, are made from them. These things force us to two extraordinary conclusions. First, the enriched gas returned to the interstellar medium must in time be used to form later generations of stars. And second, our own bodies are largely, if not entirely, made of matter that has at one time or another been part of a massive star.

After the supernova, the core of the original massive star remains as a *neutron star*, a ball of neutrons only 10 or 20 kilometres across, with a mass rather larger than the Sun. This implies a mass density of about 10^{18} kilograms per cubic metre, a truly astounding figure. We saw earlier that a teaspoonful of white dwarf matter weighed about a ton. But the same teaspoonful of neutron star matter would weigh a *billion* tons. This density is actually the same as in the nucleus of an atom, where neutrons (and protons) are packed as tightly as possible. The only reason that we are largely unaware of this is that, as we noted earlier, most of an atom is empty space, with its constituents just the 'flies in the cathedral'. A neutron star in fact resembles a giant nucleus, actually held together by gravity rather than the strong nuclear force.

In many ways, a neutron star is a vastly more compact version of a white dwarf, with the fundamental difference that its pressure arises from degenerate neutrons, not degenerate electrons. One can show that the ratio of the two stellar radii, with white dwarfs about one thousand times bigger than the 10 kilometres of a neutron star, is actually just the ratio of neutron to electron mass. So is there a maximum mass for neutron stars, like the Chandrasekhar mass for white dwarfs, which arises when the neutrons have to move at speeds near light to hold up the star's weight? Its value is more complicated to work out, but is somewhat larger than the Chandrasekhar value, probably around three solar masses. So massive stars with cores lighter than this end their lives as neutron stars. This is the second possible endpoint of stellar evolution.

Now we face another crisis like the one that the discovery of the Chandrasekhar mass caused. Can all massive stars shed enough mass to reduce their cores below this new maximum value for neutron stars? Not surprisingly, the answer to this is again 'no', just as for white dwarfs. And this time, there is no new state of matter like a neutron gas to save us from what to many astronomers seemed an unpleasant or even absurd conclusion: sufficiently massive stars cannot end their lives as either white dwarfs or neutron stars. No form of pressure can hold them up against their own weight. They must go on collapsing!

It took astronomers a long time to understand what the last statement would really mean for astronomy. When the answer emerged in the 1960s, it turned out to have more to do with understanding the mathematics of Einstein's general theory of relativity than with direct observation. This theory, usually referred to as GR (for general relativity) was proposed by Einstein in 1915, and incorporates gravity as part of the fabric of space and time rather than a force like the electromagnetic and nuclear interactions. Within a few months, Karl Schwarzschild showed what GR would imply for the space outside an isolated star (sadly dying of disease in the First World War before his paper was published). This 'Schwarzschild solution' showed that at large distances from the star, things would behave just as Newton's much older theory said they should. Einstein had, of course, deliberately constructed his theory with this property, since the older theory was known to reproduce observations within extreme accuracy in such cases. Schwarzschild's solution predicted a strange deviation from Newtonian theory for stars whose radii were very small for their masses. This was not understood at the time, nor for many years afterwards, as we shall see. There was little urgency in finding out, as the critical size, the so-called Schwarzschild radius, was directly proportional to the star's mass, and had a value of about three kilometres if this mass was the same as the Sun's. In other words, the star had to be extremely condensed indeed for whatever strange effects occurred at its Schwarzschild radius to matter for astronomy.

Since no stars with such extreme density were known or even contemplated in the early 20th century, this odd feature of the Schwarzschild solution remained something of an enigma, some physicists even attempting to remove what they saw as an unphysical blemish on GR. The theory itself became a backwater, as almost none of its more extreme predictions could be tested by experiment or observation. But by the late 1950s, the possible existence of neutron stars was being taken seriously, and these stars would be only a little larger than their Schwarzschild radii. With only simple mathematical manoeuvres (one of them ironically already made by Eddington in the 1920s, though he did not appreciate its significance), physicists finally penetrated the mystery: light from the surface of a star smaller than its Schwarzschild radius could not reach the rest of the Universe outside.

This is the ultimate triumph of the famous $E = mc^2$. Light has energy, so light has weight, and gravity can bend light. If the star is dense enough, gravity can prevent the light reaching the rest of the Universe. Applied to our problem of an ever-collapsing star, this interpretation tells astronomers not to worry. No physical effect can travel faster than light, so once a star has collapsed inside its Schwarzschild radius, no new information about its behaviour can reach or influence the Universe outside. Nothing that the collapsing star does inside this radius can possibly matter to astronomers outside it. So it is effectively a new and very strange kind of star. This is a *black hole*, and is the third of the three possible endpoints for the life of a star.

Of course, the black hole still has a gravitational field, and it continues to affect the orbits of matter around it. But in an extraordinary way, a black hole is far simpler even than a star. A normal star's gravitational field is mainly characterized by its total mass. But a real star is never quite perfectly spherical – if it rotates, it becomes fatter around its equator. If it has magnetic fields, these can alter the distribution of gas within it in complex ways. All of these things affect its gravitational field to some degree, and could in principle be detected by observations, for example of objects in orbit around it. So the gravitational field of a normal star is, at least in detail, quite a messy thing.

But the gravitational field of a black hole is amazingly simple in comparison. Of course, it is mainly characterized by the black hole's mass. But apart from this, it bears the imprint of just *one* aspect of the star that collapsed: it 'knows' about the rotation of the star (technically speaking, its angular momentum), and exerts a maelstrom-like effect on the orbits of matter very close to it. But all other properties of its stellar progenitor disappear: every single black hole in the Universe is accurately characterized by just its mass and rotation. The exact solution of the GR equations for a steadily rotating black hole was found by Roy Kerr in 1963, and this solution describes every black hole in the Universe by just two numbers specifying mass and angular momentum. Chandrasekhar's reaction to this was:

> In my entire scientific life, extending over forty-five years, the most shattering experience has been the realization that an exact solution of Einstein's equations of general relativity, discovered by the New Zealand mathematician, Roy Kerr, provides the absolutely exact representation of untold numbers of massive black holes that populate the universe. This shuddering before the beautiful, this incredible fact that a discovery motivated by a search after the beautiful in mathematics should find its exact replica in Nature, persuades me to say that beauty is that to which the human mind responds at its deepest and most profound.

How does this astounding simplicity appear in practice? After all, the star that collapsed to make the hole must have had all sorts of deviations from the pure symmetries assumed in Kerr's solution. Why do these not leave an imprint on the hole's gravitational field? The answer comes from another property of Einstein's GR: strong and rapidly changing gravitational fields spread out their irregularities in *waves*. So imagine that the collapsing neutron core of a massive star had a mountain somewhere on its surface. The mountain would cause a massive burst of gravitational wave emission, which would take gravitational energy away from the thing causing it – the mountain – making this rapidly flatten out. The timescale for all this action is just the time the core takes to collapse through its Schwarzschild radius, which it does at a speed close to light. For a core of a few solar masses, this timescale is just a few times 10^{-5} seconds – 10 milliseconds or so. The same holds for any other deviation from the pure symmetries of the Kerr solution. Black holes rapidly become the most perfectly simple objects in the Universe.

Chapter 6
Finding the bodies

The theory of stellar evolution tells us that stars can end their lives in just one of three possible ways: white dwarf, neutron star, or black hole. We saw that white dwarfs were directly observable. But neutron stars are so tiny – one million times smaller in surface area – that if they had a similar temperature, they would be extremely faint and difficult to see. And by definition, black holes do not radiate at all. Can we be sure these strange objects really exist? We will see that the evidence that they do is overwhelming. On the way, we will find that black holes in particular are a major influence in shaping the structures we see in the Universe around us.

Pulsars

The first detailed evidence that neutron stars exist emerged from an experiment designed to look for something else. Radio astronomy had become established after the Second World War, and most of the early sources were distant galaxies. Astronomers realized that the radio signals they detected must be affected by passing through tenuous gas blown out by the Sun (the 'solar wind'). This gas is easily deflected by the magnetic fields of the planets, and so can move around rapidly. This did not matter much for observing distant galaxies, because adding up the radio emission for a long time averaged out the interfering effects of the interplanetary gas. But a radio telescope sensitive enough to detect a distant source in a very short observation seconds would be able to see this interplanetary scintillation and allow astronomers to study it.

Importantly, this was the first time that astronomers had deliberately set out to make observations of such rapid phenomena. Astronomy seemed to be a subject where significant changes occurred only over millions of years, and even planetary orbits were measured in months or years. Certainly no-one predicted what happened next. In 1967, Jocelyn Bell, a graduate student in Cambridge, was monitoring the output from an instrument designed to study interplanetary scintillation when she noticed what appeared to be regular pulses of radio emission about once per second coming from a fixed point in the sky. The fixed position meant that it must be an astronomical rather than terrestrial or planetary object, but the extreme rapidity and regularity was unprecedented. After some confusion, clarity emerged.

The most obvious way to make a very regular signal is a rotation. If you paint a spot on a ball and spin it, you see the spot at regular intervals. The same works for a spot on a star, as the star spins. But there is a limit to how fast the star can spin – it deforms once the centrifugal repulsion at its equator is stronger than gravity. For a smaller star of the same mass, this limiting speed increases, since gravity at the equator increases. So the more condensed the star is, the more rapidly the spot can reappear. The one-second period of Jocelyn Bell's object required this *pulsar* (as these objects were soon named) which produced it to have a density significantly higher even than a white dwarf, and the limit quickly went still higher as more pulsars were found. Antony Hewish, the supervisor of Jocelyn Bell's PhD thesis, shared the Nobel Prize in 1974 for the discovery of pulsars.

Although the idea of a neutron star had been envisaged theoretically since the 1930s, few astronomers knew or remembered it. But the huge density required of pulsars rapidly revived interest. Theorists soon realized that a neutron star was likely to have a strong magnetic field, as even a weak field is hugely magnified by the intense squeezing involved in reaching neutron star densities. Allied with the rapid rotation, this can produce a beam of radio emission anchored to a fixed point on the neutron star (Figure 10). This beam acts like the spot on the ball we considered above, and explains why pulsars pulse. Not only are pulsars neutron stars, they are stars that shine in an entirely new way: not from internal nuclear reactions, like most stars, or even from leftover heat, like white dwarfs. The power source for pulsars is their rotation. As pulsars lose energy by radiating, their rotation must slow down, that is, the period between pulses must increase. Of course, this increase is extremely slow compared with each pulse, so we do not see the pulses getting slower before our eyes. But it is easy to test for such an increase once you know it might be there – the pulse period is so short that you accumulate thousands of pulses per day. That means you can measure the average time between pulses – the period – very precisely in only a few hours. Comparing this measurement with later ones allows you to find extremely small changes, and indeed the *rate* at which the spin of the pulsar is slowing down. Astronomers alway find that the pulsar spindown gets slower and slower over time, so that a slow pulsar takes longer to increase its pulse period by a certain fraction. Conversely, a fast pulsar spins down fast, and must have started to pulse only relatively recently. We can estimate how recently simply by dividing the pulse period by the rate at which it is slowing down. The resulting number is a measure of the pulsar's age.

Very soon after the discovery of the first pulsar, another one was found in the middle of the famous Crab Nebula, which is the remnant of a supernova (Figure 9). The pulsar is fast (30 pulses per second) and the age estimate is very short, of order a thousand

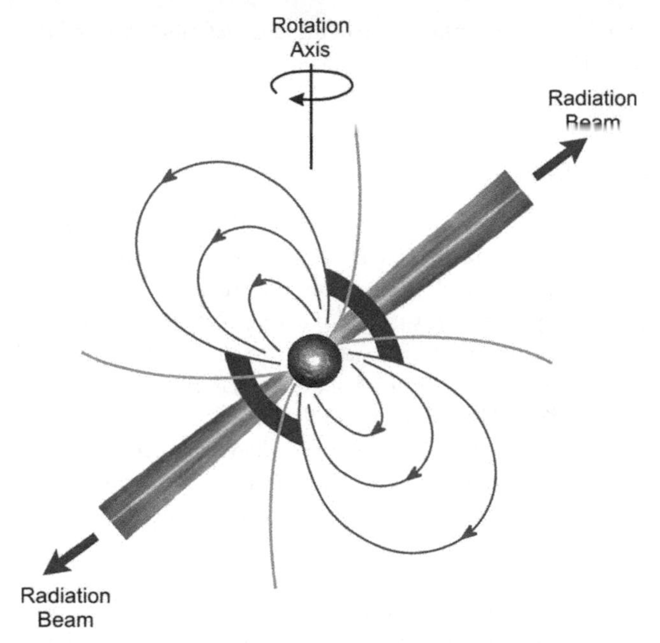

10. Schematic picture of a radio pulsar. The magnetic field lines are symmetrical about an axis which is inclined with respect to the rotation axis of the neutron star. The combination of rotation and magnetic field creates radio hotspots near the magnetic poles, and so cause the radio emission to pulse as the star spins

years. Chinese astronomers recorded the explosion of the Crab supernova in AD 1054. Although this produced a luminous patch in the sky as bright as Venus, this whole cataclysmic event was apparently completely missed in the West. The Chinese observations are crucial: the close similarity between this true age, and the age estimate from pulsar spindown, leaves no room for doubt that pulsars are spinning neutron stars, and are born in supernovae. The cantankerous Swiss-American astronomer Fritz Zwicky had already suggested supernovae as the birth sites of neutron stars back in the 1930s. He lived to see this vindication, but not the later confirmation of his idea that most of the matter in the Universe is dark, and so undetectable except through its gravitational effects.

Accretion

Even before the discovery of pulsars, the idea of neutron stars and black holes as realistic objects had been gaining ground because of another feature, which is ultimately more fundamental. This is the huge gravity near their surfaces. If matter could be dropped from rest at some distance from a neutron star, this intense force would make it fall at about one-third of the speed of light by the time it hit the surface. The collision with the surface would release all this energy as radiation. Dropping just one kilogram of matter on to a neutron star in this way releases an enormous 10^{16} joules, about 20 to 30 times *more* than getting the same mass of hydrogen to fuse to make helium.

There is a natural way of arranging that gas might fall on to a neutron star in this way. Most massive stars are not isolated, but part of a binary system, two stars orbiting each other. If one is a normal star, and the other a neutron star, and the binary is not very wide, there are ways for gas to fall from the normal star on to the neutron star. We will discuss these in detail later, but we first need to look more closely at how the gas falls.

Naturally enough, gas tends to fall most readily from the part of the normal star which is closest to the neutron star, where its gravitational pull is strongest. But since the normal star is in orbit with the neutron star, the gas falling off it is also in orbit. If only one gas particle fell towards the neutron star, it would follow an elliptical orbit, taking it quite close to that star. Because the neutron star is so tiny, the gas particle would miss it, and continue around its elliptical orbit. But in reality, we do not have just one gas particle, but a whole stream of gas particles all trying to follow the same orbit. The gas stream misses the neutron star, but as it loops back towards the normal star, the leading part of the stream collides with parts of the stream which have only just left that star. In the collision, a lot of the gas's kinetic energy, associated with its orbital motion, gets turned into heat and radiated into space.

But an internal collision of this kind cannot remove the rotation of the gas around the neutron star – this could only happen if the collision was head-on, between equal masses of gas rotating in opposite senses. The result is that the gas ends up on a circular orbit around the neutron star. Internal dissipation within the gas now means that the gas spreads into a disc around the neutron star, lying in the orbital plane of the binary. Within this disc, most of the gas is slowly spiralling inwards. This takes it closer to the neutron star, and means that the gas is losing gravitational energy. There is only one place for this energy to go – it is dissipated as heat, and then radiated into space from the two faces of this disc of gas. By the time the gas reaches the neutron star, it has released as radiation exactly one-half of the gravitational energy it would have lost by falling directly on to this star – still an enormous amount. The other half of this gravitational energy has gone into its rotation – this has been *speeding up* as the gas fell inwards, just as the virial theorem we met in Chapter 2 says. Once the gas joins the neutron star, it gives up this energy to radiation too.

Astronomers call the spiralling disc of gas an *accretion disc*, since the neutron star gradually accretes – gains mass – in this process (Figure 11). Accretion on to very compact objects like neutron stars almost always occurs through a disc, since the gas that falls in always has some rotation. This is obviously true of accretion on to a black hole, which is even smaller than a neutron star of similar mass. This fact now shows us how astronomers might do what seems impossible, and actually observe black holes. For if we replace the neutron star in the last paragraph with a black hole, the only thing that changes is what happens to the gas as it reaches the middle of the disc. It now simply falls into the hole,

11. Schematic picture of an accretion disc. Gas is pulled from the companion star by the gravity of the compact star (a neutron star or black hole in an X-ray binary system). The gas cannot fall directly on to this star, but instead orbits it in a disc-like configuration, slowly spiralling inwards on to it. This releases gravitational energy and so makes the disc shine and produce X-rays

rather than hitting the hard surface of a neutron star. But remember, the gas has already released one-half of its available gravitational energy in reaching this point, and this is a huge amount. In fact, this one-half is already bigger than the same one-half for a neutron star, since a black hole is somewhat smaller for the same mass, allowing gas to spiral into an even stronger gravitational field. So accretion on to a black hole ends up with about the same enormous efficiency in converting accreted mass into energy as we found for a neutron star, about 10^{16} joules for every kilogram. Accreting black holes should be detectable in the same way as accreting neutron stars.

How do astronomers detect these systems? Of course, they are bright objects, because accretion uses a fairly small amount of mass to make a lot of light. But there is a limit to how much light they can make, just as there is for a star. In Chapter 3, we noted that a star's luminosity cannot be bigger than the Eddington limit. At this limit, the pressure of the radiation balances the star's gravity at its surface, so any more luminosity blows matter off the star. The same sort of limit must apply to accretion: if this tries to make too high a luminosity, radiation pressure will tend to blow away the rest of the gas that is trying to fall in, and so reduce the luminosity until it is below the limit. Things are more complicated than for a spherical star getting its luminosity from nuclear burning, but roughly speaking, accreting stars obey the same Eddington limit, so again, there is a maximum luminosity they can have, specified by their masses.

So if the sheer brightness of an accreting neutron star or black hole does not pick it out from bright examples of normal, nuclear-burning stars, what does? The answer is *size*. The accretion luminosity of these objects is high not simply because their masses are large, but because their physical sizes are so small that gas can fall very deep into their gravitational fields without being stopped by their surfaces, and so release huge amounts of gravitational energy. But this means that a lot of energy must come from a very small region: a neutron star is only 10 kilometres in radius, compared with the 700,000 kilometres of the Sun. This can only happen if this very small surface gets very hot. The surface of a heathily accreting neutron star reaches about 10 million degrees, compared with the 6,000 or so of the Sun. The same temperature must hold for the inner part of the accretion disc around the star, which emits a similar luminosity from a comparable area. The radiation from such intensely hot surfaces comes out at much shorter wavelengths than the visible emission from the Sun – the surfaces of a neutron star and its accretion disc emit photons that are much more energetic than those of visible light. Accreting neutron stars and black holes make *X-rays*.

Until the 1960s, astronomers had not attempted to observe the sky in X-rays. There is a formidable obstacle to doing this, because the Earth's atmosphere absorbs X-rays very efficiently. To detect X-rays, astronomers have to fly their instruments above the atmosphere. This means using balloons, rockets, or best of all, artificial satellites. So X-ray astronomy only really got going once the space race of the 1960s made the launching of scientific satellites relatively routine. The effect was dramatic – hundreds of X-ray sources, far brighter than the corona of the Sun, which was the only known source before this. But the detected brightness does not by itself tell us the luminous output of a source: is it a brilliant object relatively far away, or a comparatively faint object which happens to be near us? As we have seen, astronomers did have a good idea of the distances to stars. So when they found a new X-ray source, astronomers looked for a visible star which could be physically associated with it. This is harder than it sounds, because early X-ray detectors could not specify the positions of sources in the sky with great precision. But X-ray sources are often seen to *vary* – to change their luminosities a lot in a very short time – because they are very small objects. If one of the visible stars close to the possible position of the X-ray source varies in time in precisely the same way at the same time, it is a pretty good bet that the visible star is physically associated with it, and in particular, at the same distance.

In this and various other ways, astronomers now have a good idea of how distant, and so how bright, most X-ray sources are. The sources divide into families. All the early discoveries were *high-mass X-ray binaries*. These are binaries systems in which a neutron star accretes gas from a very blue companion star. We know that these stars have to have high masses and luminosities. The companion completely outshines any optical emission from the accreting neutron star or black hole, which we detect only in

X-rays. In many cases, the X-ray luminosity is close to the Eddington limit for a star of about the mass of the Sun, just as we would expect if it is a neutron star or a black hole. Several of these high-mass systems give striking proof that the accretor is indeed a neutron star, because their X-ray emission repeats in a completely regular way, with a period of a few seconds. This is so similar to the periodic radio emission of pulsars that it is natural to try the same picture of them as strongly magnetic neutron stars here too. The physical difference is that the energy source for pulsing X-ray sources is accretion, not rotation. To explain why these sources pulse, we have to arrange for accretion spots on the surface rather than the spots of radio emission in pulsars. The obvious answer is again the intense magnetic field we expect a neutron star to have. This is so strong that the accreting gas cannot cross the magnetic field lines. Instead, the gas slides down these lines, landing near the magnetic poles of the neutron star, and making the accretion spots that lead to the pulsing.

The second prominent family of X-ray sources in the Milky Way are *low-mass X-ray binaries*. These systems have red companion stars, which must be low-mass objects. None of these sources show pulsing, but many of them have a new property. At intervals of anywhere between a few hours to months, they show very large bursts of X-rays, which rise rapidly to the Eddington luminosity and then decay. The X-ray spectrum of each of these bursts looks like that from a hot object, with a temperature of about ten million degrees. As the burst decays away, this temperature drops. If we know the temperature and luminosity of a hot body, we can work out its emitting area ($A = L/\sigma T^4$). As the burst decays, this area comes out at a constant value of a few hundred square kilometres. This is just the area we would expect for a neutron star, which is a ball with a radius of 10 kilometres. So something is making the surface of the neutron star hotter for a short time. This tells us that these bursts result from a new energy source. So far we have assumed that all the X-rays come from accretion, because that is the most efficient way of using matter to make radiation. However, the conditions on the surface of a neutron star are ideal for nuclear reactions. If the burning happened all the time, it would increase the X-rays only by a small amount, a few per cent, and would not be noticeable. But if the burning instead occurs in short bursts, involving matter that has accreted over the interval since the last burst, this can stand out against the steady background of X-rays from accretion. What seems to happen is that the accreted hydrogen burns steadily, and so is not noticeable, but helium burning is sporadic, and provides the bursts we see. It is very striking that X-ray sources that pulse do not burst, and those that burst do not pulse. This is probably because the magnetic field that must be present in a pulsing source actively prevents nuclear burning.

Black holes

Black holes have so far appeared in this chapter only as possible X-ray sources. It is clear that they can neither pulse nor burst, since pulses require a magnetic field and bursts need a hard surface. But this does not pick them out – there is no reason why a non-magnetic neutron star should pulse, and bursting only occurs under special circumstances. So how can we identify the black holes among the X-ray sources that neither pulse nor burst? There is only one distinctive property left: mass. Neutron stars, like white dwarfs, cannot exist above a certain mass, as we saw in the last chapter. The precise value of this maximum mass is more uncertain than for white dwarfs, because the physics determining the relation between pressure and density is complicated by nuclear physics. But simple arguments tell us that the maximum mass cannot be much larger than about three times the mass of the Sun, whatever this physics requires. So astronomers have a relatively simple way of picking out black holes from other other X-ray sources: if they can measure the mass of the accreting star and find that it is larger than three solar masses, it must be a black hole.

This sounds fine enough, but how do we measure the mass of an accreting star? The good news is that they are all in binary systems. So, as we discussed in Chapter 1, we can try to measure their masses by seeing how fast they fling their companion stars around in orbit. We need to take spectra of the companions, and see how these move over time. Then the Doppler effect tells us something about the speed – that is, the back-and-forth motion along our line of sight. As we saw in Chapter 1, this gives us a rock-bottom lower limit to the mass of the X-ray star.

This all sounds very straightforward, but there is a big snag. X-ray binaries are distant objects, so they appear as just one object in a telescope – we cannot separate out the companion star and take its spectrum alone. Worse, the X-ray source is surrounded by an accretion disc, which produces a lot of optical light. Often, this drowns out the faint spectrum of the companion star, making it impossible to measure this to get speeds and so a mass.

Fortunately, nature makes it relatively easy for astronomers in one class of X-ray binaries. These are the *transient* sources. These systems spend most of their lives not producing X-rays at all. But from time to time, they do make X-rays, and indeed get very bright, showing that indeed they really are binaries which must contain either a neutron star or a black hole, before they eventually switch off again. We now know that this apparently capricious behaviour happens because the accretion disc around the (occasional) X-ray star is unstable. Most of the time, gas transferred from the companion star slowly accumulates in this disc without much of it ever getting near to the neutron star or black hole in the middle. The disc stays very cool and faint because not much gas is getting deep down into the gravitational field. But eventually, the disc cannot stay in this quiescent state: it heats up, and in doing so suddenly forces a lot of gas to fall inwards, making more X-rays. These in turn heat the disc, and make more gas fall inwards, making the system brighter still.

This kind of runaway cannot go on forever: gas falls on to the central accreting star faster than it is transferred from the companion star, so the disc will eventually run out of it. The disc may be able to cool down and stop the accretion even before this happens. The outbursts can be quite long, months or years, because the disc can hold a lot of gas. The faint quiescent phases may last for decades or more, because refilling the disc with gas from the companion star is a slow process. Unless astronomers have seen an outburst from the system in the past, it is so faint that it is almost impossible to discover in this phase. There must be a lot of systems like this which have not yet been seen to have an outburst, and will remain undiscovered for decades or even centuries until they do.

These alternating bright and faint phases are ideal for estimating the mass of the accretor. The bright X-ray phase tells astronomers that there is a neutron star or black hole in the binary. When the X-rays fade, astronomers know that there will be a faint optical source in the same position, which is now dominated by the light from the companion. They can then get clean spectra of this star, and find the orbital period by waiting for these to repeat. Knowing this period and the total change in the line-of-sight speed gives the rock-bottom mass estimate we mentioned earlier. Astronomers have found more than half a dozen systems for which the rock-bottom mass is bigger than three solar masses, making a black hole the only possibility. A longer list of more than 20 systems appear to have black holes if we make plausible guesses about the angle of the binary orbit to our line of sight, and the mass of the companion star. It seems that black holes in binaries could have masses anywhere up to 20–30 solar masses, or even more.

So the evidence that black holes with masses typical of stars exist is overwhelming. The evolution of a star ends in just one of three ways, all of them very compact: as a white dwarf, neutron star, or black hole. The star's mass at the start of its life fixes how it will end. We saw already in Chapter 5 that stars initially of seven solar masses or less end as white dwarfs. More massive ones end as neutron stars or black holes, but we do not currently know very well just what masses of stars end in each way. We would naively expect that the black holes always descend from more massive stars than neutron stars, and this seems to be largely correct, but the physics of core collapse is so complicated that even this is not settled yet.

How binaries evolve

Now we should return to an important question we bypassed near the start of this chapter. All of the stellar-mass black holes we know, and many of the neutron stars, are found in binary systems – two stars orbiting each other. We see these exotic objects only because gas falls from the companion star on to them. How does this happen? Usually the binary system starts off as just a pair of main-sequence stars born at the same time (although we will discuss an exception to this in the next chapter). Over time, each of the two stars burns hydrogen, pretty much unaffected by its companion. Eventually, the more massive star evolves off the main sequence. As we know, this means that its core shrinks and its envelope starts to expand towards a giant size. Unless the binary is very wide indeed, the surface of the expanding giant quite quickly gets close to the companion. Now it really notices the gravity of this star for the first time, and gas falls from its outer layers towards this star. Usually the companion is large enough that this gas hits its surface and directly joins it, rather than orbiting and forming one of the accretion discs we discussed earlier.

Three things now begin to happen at the same time. First, the star losing the mass has to get used to the idea – it has to adjust its internal structure, now evolving slightly more slowly and perhaps changing its radius slightly. Second, the companion gains mass, and also has to adjust, speeding up its nuclear evolution a bit, and adjusting its radius. Finally, the whole binary has to adjust. Why? The process of *mass transfer* takes gas from the more massive star, and puts it on the less massive star. But the more massive star was closer to the binary's centre of mass than its companion: the transferred mass started on this star but ends up on the companion, so on a wider orbit than before. If nothing else happened, this would mean that the total angular momentum – the amount of rotation – of the whole binary system had increased. But this cannot happen! The binary system is isolated, and has no way of gaining angular momentum. The only way out of this apparent paradox is that the extra angular momentum of the transferred mass is balanced by a decrease in the angular momentum of the rest of the system. This must mean that the distance between the centres of the two stars – the binary separation – *shrinks*.

You can easily see what will happen now: the shrinkage of the binary means that even more gas gets transferred, which shrinks the binary still more, and leads to more mass transfer. This positive feedback means that systems like this transfer mass at very high rates and can be extremely bright. Exactly the same argument in reverse tells us that transfer of mass from the less massive star in a binary to the more massive makes the system become wider: mass is ending up nearer to the binary centre of mass, so to keep the binary angular momentum constant, the binary should *expand*. Pulling the stars apart in this way must tend to shut off the mass transfer. Yet there are many binary systems like this, for example where a white dwarf gains mass from a low-mass main-sequence star, which appear to have been happily transferring mass for a very long time. This must mean that some process *removes* orbital angular momentum from the binary. How can this happen?

There are several ways, but one of them is particularly important. If the stars are close enough – binary periods of less than two hours or so, the two stars are moving quite rapidly in their orbits (several hundred kilometres per second). Einstein's general theory of relativity predicts that large masses revolving at high speed should emit gravitational waves. These are ripples in space–time which take energy away from the binary orbit, and so make it shrink. At these orbital periods, the energy loss is fast enough that the binary separation decreases on a timescale significantly shorter than the age of the Universe (about 10^{10} years). This shrinkage means that there is less room inside the binary system for the companion star to fit in. It finds itself so close to the accreting star that its gravity cannot retain the gas nearest to that star. So in the binaries I mentioned above, gas must be transferred from the low-mass main-sequence companion to the white dwarf, in step with the rate at which gravitational waves carry off angular momentum. This is faster than the nuclear evolution of the companion star, so this nuclear evolution never happens. The transferred gas forms an accretion disc around the white dwarf. This disc cannot soak up a very large mass, so all the transferred gas must end up spiralling down to the white dwarf, giving up gravitational energy to radiation as it does so. The rate given by this process seems to be about right to explain the brightness of the accretion discs in these systems that astronomers measure.

All this shows that binary evolution can get complicated – not only do the binaries expand or contract depending on which way mass is being transferred, the stars themselves can evolve, which can mean expansion, or the binary can lose angular momentum, which makes it shrink. The speed of evolution depends on the mass of the star, but mass gets exchanged. This can sometimes leave us with the apparent paradox of the less massive star being further along in its nuclear evolution. Which particular effects win out depends ultimately on the masses of the two stars and the distance between them right at the start of their evolution. So it is not surprising that a huge variety of binaries is observed.

As if this was not enough, binaries containing massive stars face a mortal threat as one or both of them undergoes a supernova explosion. A core-collapse supernova suddenly removes a large fraction of the total mass of the binary. If this fraction is more than one-half and the explosion is symmetrical in all directions, the binary inevitably breaks up. The sudden mass loss means that the two stars in the binary – the companion and the collapsing core of the supernova – have no time to adjust their speeds as the mass is lost. There is too little mass left in the binary for gravity to hold on to these two speeding stars, and they fly apart.

This is an alarming prospect: since neutron stars (definitely) and black holes (probably) are born in supernova explosions, this suggests that it might be difficult for them to remain in binaries at all. Yet binaries are the only places we can be sure that we see stellar-mass black holes, and there are many binaries containing neutron stars. So evidently at least some binaries know how to survive a supernova explosion.

We can guess how this might happen from our observation above that the less massive star can appear to be more advanced in its evolution, because the binary has exchanged mass at an earlier stage. So although we would normally expect that the more massive star would reach the supernova stage first, this might sometimes be reversed. How does this help? Remember that the threat to the binary comes if it loses more than half its mass in the explosion. If the more massive star explodes, it loses a lot of its mass, which may well amount to more than the deadly half binary mass. But if instead the binary has exchanged mass earlier, and the supernova explosion takes place in the less massive star, there is a good chance that the total mass lost is less than a half of the binary total. Such binaries would remain together. This must be what happens in high-mass X-ray binaries. The originally more massive star expanded to giant dimensions and transferred its envelope to its lower-mass companion, reversing the mass ratio. The core of the giant continues its nuclear evolution and eventually makes a supernova, but this removes relatively little mass and the binary stays together. The massive companion is now extremely luminous. Stars like this lose huge amounts of mass into space simply because they are very bright, and their radiation blows gas away from their surfaces in what is called a stellar wind. The compact star is just a tiny object orbiting within this wind. It needs only to capture a minute fraction of this wind gas to turn itself into a respectable X-ray source – this is a high-mass X-ray binary.

But astronomers also observe low-mass X-ray binaries, where the remnant of the supernova (a neutron star or black hole) is by itself already more massive than the companion star. Before the supernova, the disparity in masses must have been still greater, so there is a very good chance that the explosion removed more than the killer one-half binary mass. How does the binary survive?

The answer is only rarely, and even then, essentially by good luck. The low-mass X-ray binaries that astronomers observe are rare and improbable survivors of a hazardous past. So much so, that I remember a conference speaker about to give a talk on low-mass X-ray binary evolution being introduced by the chairman with the words 'this had better be a very unlikely story'.

The good luck here seems to be that supernova explosions are not completely symmetrical in all directions, either in the amount of mass they lose, or in its speed. This lopsidedness means that the core of the supernova – the future neutron star or black hole – tends to fly off in the opposite direction to the net thrust of the explosion, rather like a rocket moving in the opposite direction from its exhaust. Just occasionally, purely by chance, the direction of this kick closely matches the way that the companion star happens to be moving in its orbit. So in most cases, the binary flies apart at high speed, but just very rarely, the neutron star or black hole chases after the companion and the binary stays together. As a natural result, this binary itself has quite a high-speed orbit in the host galaxy, which is just what astronomers observe for low-mass X-ray binaries.

The last few paragraphs show that high-mass X-ray binaries are a natural outcome of binary evolution, but low-mass ones are very rare freaks. You might imagine that this would mean that there are far more high-mass systems than low-mass ones. Yet astronomers see roughly comparable numbers of these two types of systems in the Milky Way. The way out of this puzzle is that high-mass systems have far shorter lives than low-mass ones. The latter live quiet lives, stably transferring mass either by losing angular momentum, or because the companion star slowly expands as a low-mass giant. By contrast, high-mass X-ray binaries have colourful but brief lives. After only a few hundred thousand years, the high-mass companion star starts to expand rather rapidly, and swamps the system with gas. There is so much of this that the X-rays now cannot get out, and so the system is no longer recognizable as a high-mass X-ray binary.

Endpoints: stellar recycling

So what happens to these two types of system in the end? Let's think about the low-mass systems where mass is transferred because the companion expands as a giant. Because this star is less massive than the compact component (neutron star or black hole), things are now fairly orderly. The companion has a burning shell of hydrogen around a low-mass white dwarf core. The core grows as the helium ashes rain down on it, and the red giant envelope expands. This causes mass transfer to the compact component, and the binary gets wider in step with it because mass is transferred from the less massive to the more massive star. Such systems usually end with the compact component in a very wide orbit with a white dwarf, the core of the giant. If the compact star is a black hole, this kind of system is almost impossible to observe, as there is no accretion to light this up, and the white dwarf is too faint to see at the likely distance of such a system. But if the compact component is a neutron star, something very interesting happens: in a significant number of cases, this neutron star appears as a radio pulsar. And more than this, the pulsar is extremely rapid, spinning with a period of only a few milliseconds.

How can this be? We started with a neutron star accreting mass from a giant companion. To end up as a pulsar, the neutron star must have some kind of magnetic field. So it must have been a pulsing X-ray source. Astronomers observe a number of these – they have spin periods typically of seconds or more, but never as short as milliseconds. From various arguments and measurements, we know that their magnetic fields are enormously strong – about 10^{12} times the Earth's magnetic field that makes compass needles point north on Earth. With such fields, the gas that accretes on to the neutron stars is controlled by the field at a large distance from their surfaces. The gas cannot cross the field lines, but must move up or down them. At spin periods of only milliseconds, this gas would be spun around by the magnetic field so fast that it would fly off centrifugally – gravity would be too weak to pull it in. There would be no accretion, and no pulsing X-ray source. Strong magnetic fields and rapid spins are incompatible for accreting neutron stars. So the millisecond periods seen in the radio pulsars in these very wide binaries mean that the magnetic fields of the neutron stars are now much weaker – about a thousand times weaker – than when they were first accreting.

Some process, which is not yet understood, has made the magnetic field decay. This probably happened at some point while gas was still falling on them and making X-rays. The fields would have become weak enough that the gas could push them aside and land directly. This gave the neutron star the rotational energy it needed to eventually appear as a radio pulsar, though of course the radio pulses astronomers observe can only appear once all accretion has totally stopped. Astronomers refer to pulsars which acquired their rotational energy this way as *recycled*. Because their magnetic fields are so weak, recycled millisecond pulsars lose only a tiny fraction of their rotational energy as they radiate. Their pulse periods remain almost unchanged however long they are observed, and make them actually more accurate timekeepers even than atomic clocks.

Endpoints: the brightest binaries

Recycled pulsars are a classic outcome of binaries where the compact accreting component is more massive than its companion star. What happens to high-mass systems? In the X-ray binary phase, the bright massive companion loses a lot of gas into space. The neutron star or black hole picks up only a tiny fraction, so there is no tendency for the binary to shrink yet – in fact, it widens slightly, because the mass loss means that there is less gravity keeping the two stars close to each other. But quite soon, the companion evolves to become a supergiant, growing its radius to the point where the compact companion's gravity directly captures gas from its outer layers. This is dangerous for the binary because mass is now being transferred rather than lost. We saw earlier that moving mass from the heavy star to the light one (neutron star or black hole) makes the binary shrink and so increases mass transfer. This positive feedback makes mass pour over at huge rates that swamp the X-rays and ends the binary's life as a high-mass X-ray binary.

The positive feedback only ends when the star cannot keep up with it. Eventually, it starts to shrink as it loses mass, so bringing the stars closer together no longer forces mass to pour over even faster. Things stabilize when the star shrinks at the same rate as the binary. But this is still an enormous rate of mass exchange, enough to transfer the entire companion star in a few hundred thousand years if it were to continue. This is far beyond what the compact accretor can accept – several thousand times its Eddington limit.

Astronomers know one system which is probably of this type, called SS433. Perhaps oddly, SS433 is not a bright X-ray source at all. Its most spectacular feature is that it produces two oppositely directed jets of gas, which precess around a fixed direction in space every 163 days. Radio astronomers can actually see these jets (Figure 12). The jets emit spectral lines, and the Doppler

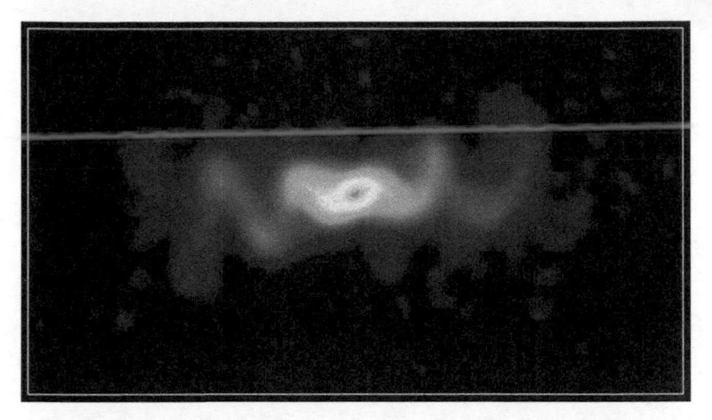

12. The precessing radio jets of the binary system SS433. The jets are produced by gas accretion on to the central black hole (or possibly neutron star), but precess in space, tracing out a corkscrew pattern on the sky

effect shifts these back and forth as the jets precess. These shifts allow astronomers to disentangle the angles and velocities involved. The jets precess at about 23 degrees to a fixed axis, which is at about 79 degrees to our line of sight, so lying almost in the plane of the sky. The line emission is actually characteristic of hydrogen, but hugely red- and blue-shifted because of the enormous velocity of the gas in the jets. This is moving out at about one-quarter of the speed of light. This last fact strongly suggests that we are dealing with accretion on to either a neutron star or a black hole, because the maximum speed of outflows is generally similar to the speed of infall on to the star concerned. SS433's other distinctive feature is that it is losing mass in a spherical outflow at an enormous rate, about a solar mass every hundred thousand years. This is far more than even a black hole of any reasonable mass (less than one thousand times the Sun!) could accept. So this mass outflow rate must be only slightly less than the rate at which mass is transferred from the companion star in the binary.

All of these things add up to a picture of a hugely energetic object. The mass transfer rate we deduced above is just about what we expect if a massive star is transferring mass to a lighter black hole or neutron star. It forms an accretion disc around this object, but most of the mass is ejected from the disc by radiation pressure, as the accretor desperately tries to keep the accretion rate below the Eddington limit at each radius. It comes out more or less spherically, except near the middle of the disc. Here it must leave a pair of empty funnels, because gas moving along the disc axis would have no angular momentum at all, and the gas reaching the inner edge of the disc finds it hard to lose any. Only a small part of the gas manages this, producing the jets.

These funnels are important. Radiation produced by gas spiralling in through the disc cannot immediately escape because of all the gas being ejected too. The radiation is scattered in random directions by this gas, and eventually finds its way to the open funnels, where it is able to escape freely. This explains why SS433 is only a weak source of X-rays, since these must come out along the funnels, which permanently point away from our line of sight.

On the other hand, anyone viewing a source like this from directions within one of the funnels would see an unusually bright object, since all of the accretion luminosity is crowded into these two narrow cones instead of being spread out in all directions. This may be the explanation for the ultraluminous X-ray sources, objects whose X-ray flux is so high that if they were radiating this much in all directions, their luminosities would be far above the Eddington limit for a stellar-mass black hole. These objects are all in fairly distant galaxies, and there is so far none known in our own galaxy, the Milky Way. This is reasonable: the narrow ultraluminous funnels of these systems point in totally random directions. Only a tiny fraction of them can be pointing directly towards us, so that we receive radiation from the funnels and see them as ultraluminous X-ray sources. To find this tiny fraction means searching through a large number of such objects, most of them pointing elsewhere and not recognizable to us, and spread over a huge volume of space. It is overwhelmingly likely that the nearest system arranged with its funnels pointing towards us is very distant. Astronomers have now found a few dozen, all in distant galaxies.

The ultraluminous stage is not the endpoint for high-mass X-ray binaries. The massive companion star continues to evolve internally even after its hydrogen envelope has been accreted by the black hole or neutron star. Eventually, it explodes as a supernova. There is a good chance that the binary can survive this, as the exploding star has a low mass. Depending on what the accretor was, and what the supernova leaves behind, we get a binary system with two black holes, two neutron stars, or one of each. So far, astronomers have not found any double black hole systems – this is not surprising, as there is no mass transfer and no X-rays. But astronomers can find systems that contain neutron stars which are radio pulsars, and these are extremely important.

The reason is that we can measure the motion of a pulsar in a binary system. As the pulsar moves around the binary centre of mass, it moves alternately away and towards an observer on Earth. As it moves away, successive pulses have very slightly further to travel to reach the telescope, and so each one arrives with a slight delay. Conversely, the pulses arrive slightly early if the pulsar is moving towards us. This is exactly like the Doppler effect on the spectral lines of a normal star in a binary system, and in the same way, the pulse arrival times tell us the velocity of the pulsar along our line of sight. Newton's law of gravity now gives us a tight constraint on what masses the pulsar and its companion can have.

But things go much further. Usually, these binary systems have quite short orbital periods of only a few hours – remember that the transfer of mass within the binary was already making it shrink. The short period means that the stars move rapidly, and this is just what makes the system emit gravitational waves. As we know, this makes it lose angular momentum and shrink the binary orbit, making the orbital period decrease. Here the effect is stronger because there is no mass transfer trying to widen the binary. Astronomers can detect this decrease of the orbital period very precisely in several systems. The first of these binary pulsars was discovered by Russell Hulse and his PhD supervisor Joe Taylor, who shared the 1993 Nobel Prize for this achievement. In every case, the observed decrease agrees with the one predicted by Einstein's general theory of relativity to within this very precise accuracy. Einstein's theory predicts a host of other effects, which can also be measured – the binary orbit is usually elliptical, and should precess at a certain rate. So far, none of the many measurements has turned up a discrepancy with the predictions of Einstein's theory. As well as this, these measurements give the masses of the pulsars and their companions with a precision unknown in any other field of astronomy – currently to about four significant figures.

Gamma-ray bursts

There is a spectacular phenomenon probably associated with stellar evolution which is not yet understood. Bursts of very energetic electromagnetic radiation – gamma-rays – appear at random points in the sky without warning. The distribution of these bursts on the sky does not single out any special direction, and we now know that this is because they are extremely distant. Typically, their light has been travelling to us for half the age of the Universe or more (in the language of Chapter 7, they are typically at redshifts 1 and higher). This huge distance, together with their observed radiation flux, must mean that they are extremely luminous. If they emitted equally in all directions, the luminosity of a burst would exceed the total light emitted by *all of the other contents of the Universe* for a few seconds. The total energy involved in producing such a burst would be roughly equivalent to turning the whole of a mass like the Sun's into pure energy, in a few seconds.

It appears from several lines of argument that these bursts probably do not emit equally in all directions, but in a beam, like a car headlight. This means that we only see those bursts pointing towards us, and there must be many we miss. In return, the total luminosity and energy of these bursts is considerably smaller than suggested in the last paragraph, but still enormous. It seems that the longer bursts (more than a few seconds) may be particularly energetic supernovae (hypernovae), perhaps representing the birth of a black hole. What picks them out may be that they were rotating rapidly before the explosion occurred, possible because the star was in a tight binary system. The shorter bursts (less than a few seconds) may well be signalling the coalescence of a pair of neutron stars as gravitational radiation removes their orbital angular momentum. These would then be the endpoints of the evolution of a massive but close binary.

Epilogue

We have travelled a long way in this chapter. But the concepts we have discussed here go still wider. The whole of modern astronomy is profoundly influenced by the concepts of black holes and accretion on to them. This is the most efficient way of getting energy from ordinary matter in the Universe – only the direct annihilation of matter and anti-matter can produce more, and this is not a practical way of powering realistic astronomical systems. The ideas developed here are fundamental in explaining the most energetic objects we know of, and their relation to the galaxies they inhabit. Astronomers now believe that the centre of all but the smallest galaxies contains a black hole with a mass many times the Sun's, sometimes a billion times or more. These huge black holes probably grow by accreting matter, and appear as the nuclei of quasars when they do. There appears to be a close connection between the masses estimated for these holes, and properties of their host galaxies. A natural explanation is that the holes have somehow influenced the way that the galaxies themselves were assembled. What makes this idea plausible is that, as we know, matter gives up about 10% of its rest-mass energy as it accretes on to a hole. This binding energy is more than enough to affect the entire galaxy.

Chapter 7
Measuring the Universe

You will probably agree by now that, although some puzzles remain, we have in general a good understanding of how stars evolve. Simply applying the laws of physics to a self-gravitating ball of gas reproduces much of what we can see, in amazing detail. From this primitive starting point, we have ended up understanding the chemistry that makes life, how stars live and die, and how to find black holes; and a thousand other details too.

But scientists are never satisfied: in Hobbes's phrase, 'there can be no contentment but in proceeding'. If we understand something, we want to use this knowledge to understand something else. In this sense, scientists are like viewers of long-running crime dramas – they want to find out who did it, but do not want the drama to end. And just as in crime dramas, the best moments come when two things that we think we understand don't seem to fit together. In science, this kind of conflict means that we do not actually understand at least one of these things as well as we thought, so we learn something when we finally do.

Stars are not distributed evenly over space, but are piled up in galaxies, which are themselves generally well separated from each other. Knowing how stars work means we know how old they are, and how bright they are. The obvious next step is to use this to work out how old galaxies are, and how far away they are. We will see that the answers are interesting.

Age

Bright blue stars are young. The Hertzsprung–Russell (HR) diagram shows that all these stars are on or near the main sequence and have masses ten or more times that of the Sun. Once they have evolved off the main sequence, they are never blue and bright again for any extended time. The main-sequence lifetime of a ten solar-mass star is only one-hundredth of the Sun's – a hundred million years. So left to itself, a collection of stars all born at about the same time automatically becomes redder. Collections like this exist, and are called globular clusters. They typically contain about a hundred thousand stars, and their HR diagrams are quite distinctive: the main sequence effectively stops at a fixed mass or luminosity. All the stars brighter than this turnoff are on the giant branch. Generally, the turnoff point corresponds to a mass somewhat less than the Sun, which tells us that all the stars in the cluster are older than the Sun's main-sequence lifetime, that is, more than ten billion years. Clearly, the cluster, and the galaxy, and the Universe, must all be older than this.

But even these ancient objects are not the first stars, because their spectra tell us that they contain metals. We know that elements beyond helium are made in massive stars, so globular cluster stars must have formed from gas expelled by supernovae from a still more primeval population of massive stars. This does not increase our estimate of the age of the Universe by a lot, since these stars had such short lifetimes. When astronomers do these calculations as carefully as they can, the oldest stars in the Milky Way come out with ages of about 12.5 billion years, with an uncertainty of about 1.5 billion years either way.

By itself, this number just seems like a long time. Like all such numbers, it gets interesting only when we start to compare it with others we know. First, it tells us that the Sun is about half as old as the Milky Way, reinforcing the impression that there is nothing special about our star except that we live near it. But things get really interesting when we try to compare this age with that of the entire Universe.

Astronomers have known since 1929 that we live in an expanding universe. In that year, Edwin Hubble published data showing that all distant galaxies recede from us, and that the speed of the recession is just a fixed number (Hubble's constant) times the galaxy's distance from us. Since there is nothing special about our own galaxy compared with any other, this must be true of all galaxies: like spots on an inflating toy balloon, they all get further away from each other as the balloon (or the Universe) expands. Unless we live in a unique moment in the history of the Universe, all galaxies must have been ever closer together in the past. If you know how far apart any pair are now, and how fast they are moving away from each other, you know that they must have been extremely close together at a time in the past given by dividing their distance by that speed. Hubble's law relating distance and recession velocity tells us that this time is the same for any pair of galaxies we choose, and is just the reciprocal of Hubble's constant (i.e. we divide 1 by the value of this constant). Astronomers call this the Hubble time, and it obviously gives an estimate of the time that our Universe has existed in anything like its present form from its origin in what we call the Big Bang. And now we reach the potential conflict which makes such things exciting – Hubble's law uses *entirely different* observations from those giving us the ages of stars. There is no reason for the these two enormous timescales to know about each other. But for stars to be older than the Universe would be absurd. Is the Hubble time longer than the age of these stars?

Distance

The easy part in finding Hubble's constant from a sample of distant galaxies is the speed of recession – all we need here is a spectrum of each galaxy. These are an amalgam of the spectra of all the stars in the galaxy, from which we can measure a Doppler (red)shift. This just means that the spectrum must be good enough that the observer can identify the most prominent lines (usually hydrogen) and compare their wavelengths with the laboratory value. This tells us the speed with which the galaxy is receding. For distant galaxies, this speed is far greater than the speed of the stars within the galaxy, so although these broaden the spectral line because there is a range of stellar speed contributing to the total line feature, this does not prevent us measuring the whole profile if shifted bodily to longer wavelengths. The hard part is measuring the distance. This involves identifying something whose luminosity you think you know in absolute terms (often called a 'standard candle'), and comparing this with the faint brightness you see from it. Because light spreads out as the square of the distance from the source, this comparison tells you how far away the object is. Clearly, you want your standard candle to be as intrinsically bright as possible, so that you can still see enough light from it to measure when its host galaxy is very distant. In 1929, Hubble used the brightest stars in a galaxy, and the light of entire galaxies, but this is not a very accurate method. Very bright stars are rare, so it is chancy just how bright the brightest one in a galaxy is. Galaxies can contain very different numbers of stars, and if they are disc galaxies, can have different angles to our line of sight, which will not be obvious if they are very dim and distant.

Modern telescopes and detectors are now so much more sensitive than the equipment Hubble had available that one can directly measure the Hubble constant far more accurately now. The best standard candles are stars whose luminosities vary in time in a regular way, called Cepheids. These are bright stars lying in a specific region of the HR diagram. The variations repeat with periods ranging from about one to seventy days. Importantly, we know from Cepheids in our own and nearby galaxies that the absolute luminosity of the star depends uniquely on this period, the stars with longer periods being systematically brighter. So if astronomers can identify a Cepheid in a distant galaxy, observing its period tells them how bright it is, and comparing with its observed brightness tells them how distant the galaxy is.

For the purpose of measuring distances, and so Hubble's constant, the physical origin of the period–luminosity relation is unimportant – one can simply accept it as an empirical fact. But scientists always want to understand, and here it is fairly straightforward. The star's light changes because it expands and contracts in a regular way. This is unusual, because most stars find an equilibrium radius and stay at it. Cepheids (and a few other stars) have the peculiar property that the ability of light to escape from their cores is very sensitive to how hot they are, because this depends on whether helium within the interiors has two electrons missing or one. Unusually, the first form is better at blocking the light, but corresponds to a hotter interior. At the faintest part of a Cepheid's cycle, the gas outside the core layers is hot and opaque. The star's radiation is trapped by it, heating it, and makes it expand. As this gas expands, it cools, and becomes more transparent, allowing the radiation to pass through towards the star's surface and making the star's luminosity higher. This stops the expansion, and the star now contracts under gravity, returning to the start of the cycle. Because the whole cycle depends on the amount of helium blocking the light of the star, Cepheids behave differently depending on their chemical composition, so astronomers have to be careful to take this into account.

Where does the relation between period and brightness come from? The star's luminosity changes because its size changes. The observed periods of no more than about seventy days are far shorter than the time for thermal changes (millions of years), so the surface temperature cannot be changing very much. Now we remember that for a fixed surface temperature, the luminosity of the star is just proportional to its surface area, which goes as the square of the radius ($L \propto R^2$). So doubling the radius would increase the luminosity by a factor four, while increasing it by 25% would increase the luminosity by about 50%. The change of radius must mean that the hydrostatic balance between pressure and gravity is disturbed, so the period is just the dynamical time, or roughly radius R divided by the gravitational free-fall speed ($=\sqrt{GM/R}$). The latter scales inversely with the square root of the star's radius: a star of the same mass but four times the radius has one-half the free-fall speed, and so on. So the dynamical time, and hence the pulsation period, scales like radius *multiplied* by the square root of the radius ($P \propto \sqrt{R^3/GM}$). So a star with the same mass but four times the radius should have $4\times\sqrt{4}=4\times2=8$ times the period. And we already know it should have 16 times the luminosity. We can show that the luminosity should grow just slightly faster than the period (mathematically, as the 4/3 power, i.e. $L \propto P^{4/3}$). This is about what is observed: the longer the period, the brighter the Cepheid.

Now back to the distance measurements. Using the Cepheid method, these require large international teams of astronomers, and appropriately use the Hubble Space Telescope. The result gives an age for the Universe of about 13.6 billion years, allowing time for galaxies to have begun forming promptly after the Big Bang. Happily, these results also agree very well with entirely independent measurements using the cosmic background radiation left over from the Big Bang.

Acceleration

Although all this agreement is comforting, we should remember that every standard-candle method is limited: clearly, we can say nothing about the rate of expansion of the Universe at distances (and so redshifts) so large that the candle is too dim to see. To look further back in time, we need a brighter candle, but one that is still reliably standard. Supernovae are about 100,000 times brighter than Cepheids, so might let us measure distances larger by a factor which is the square root of this, so about 300. But core-collapse supernovae are definitely not standard candles: they have quite a large range of luminosities. It is not easy to correct for this, because we do not usually know anything about the star that makes the supernova – before the explosion, the star is generally inconspicuous, and afterwards it is too late to learn anything about it.

But there is a second kind of supernova which does seem to be more reliably constant in its luminosity. This is called Type Ia, and seems to involve a white dwarf gaining mass from a binary companion and approaching the Chandrasekhar limit. At this mass, it can no longer remain a white dwarf, and must collapse, making a supernova. Because exactly the same white dwarf mass is involved each time, these supernovae are much more similar to each other than the usual core-collapse type. The likely picture for a collapsing white dwarf is that carbon starts to burn. In a normal star supported by gas pressure, the huge increase in the temperature would raise the pressure, expanding and cooling the burning gas, so that the burning would stop. But degeneracy pressure does not care about the temperature, and so this safety-valve mechanism does not work. Instead, the carbon-burning reaction runs away, and ejects most or all of the star's mass into space.

So adding mass to a white dwarf and gently pushing it over the Chandrasekhar limit seems a promising way of explaining the uniformity of Type Ia supernovae. To complete the picture, we need to find a way of making white dwarfs gain mass. Surprisingly, this is not so easy. Just adding mass gently from a low-mass main-sequence companion star in a close binary system does not work. After the white dwarf gains quite a small mass in this way, runaway nuclear burning on its surface ejects at least this mass, if not more, in a *nova* explosion. This is much weaker than a supernova, and just reduces the white dwarf mass to the value it had before any was added, or even lower. So it is hard to reach the Chandrasekhar mass this way. To hang on to the mass, the white dwarf probably has to burn the hydrogen to helium as it arrives. The accretion rate has to be quite high for this, but not too high: beyond a certain rate, the white dwarf just behaves as if it were the core of a red giant, with a hydrogen burning shell on its surface, and a very large envelope outside: remember that we saw in Chapter 5 that red giants behave the same whatever the mass of their envelopes, because these are so large that they have almost no weight.

To make a Type Ia supernova, then, we need to make a white dwarf accrete at just the right rate, neither too low nor too high. There are two ways suggested for this. First, the white dwarf may accrete from a close companion star which is more massive. Any slight expansion of this companion makes the binary shrink, and increases the mass transfer rate. This eventually stabilizes at a high rate, fixed just by the ability of the companion star to keep on expanding thermally at a steady rate. In some binary systems, conditions may be just right to make this thermal-timescale rate land in the narrow range for steady nuclear burning as the matter lands on the white dwarf. The second way is to imagine that the white dwarf is in a tight binary with another massive white dwarf. This expands as it loses mass, making the mass transfer rate increase rapidly. This is a runaway process: the more mass is lost, the faster the white dwarf loses mass, and so on. Very quickly, the two white dwarfs effectively merge. If the total mass of the two white dwarfs is bigger than the Chandrasekhar limit, the core of the merger must begin to collapse at this value, again triggering a Type Ia supernova. As yet, there is no clear observational distinction between these two ideas.

Whatever the origin of Type Ia supernovae, using them as standard candles for the distance scale produces a spectacular result. The expansion of the Universe is now apparently *accelerating*: it is faster now than at the time when the most distant supernovae exploded. This forces a considerable revision of cosmological theory, so there has been a lot of work on whether the standard candle idea is justified. While there are deviations from a strictly constant luminosity, these do not appear to evolve systematically with cosmic time. So while Type Ia supernovae may not be perfect, there is as yet no reason to believe that they may be misleading astronomers about the cosmic acceleration.

Chapter 8
The beginning

Cycles

In this book we have followed the evolution of stars to its endpoints. But perhaps surprisingly, we have said almost nothing about how stars begin their lives. There are two reasons for this. First, it works. We found that the evolution of a star is fixed almost completely just by its mass alone. Without knowing how it formed, we were able simply to start with a star of a given mass and ask how it would maintain itself in equilibrium. As we have seen, this unrolls the whole story of its future development in a way which appears almost independent of how it came to be a star in the first place. The 'almost' here is a reminder that a star's evolution is also influenced by its initial chemical composition - less than by its mass, to be sure, but still sometimes significantly, as the differing period–luminosity relations for Cepheids of different composition show.

The second reason for leaving star formation to the end is that it is harder to understand than any other part of stellar evolution. So we use our knowledge of the later stages of stellar evolution to help us understand star formation. Working backwards in this way is a very common procedure in astronomy - for example, in studying the evolution of the whole Universe, we clearly have no other choice.

Stellar evolution gives us a number of clues as to how stars form. First, we know that the chemical elements beyond helium are all made in stars. And we know that stars – even with the same mass – can have varying amounts of these heavier elements. Since massive stars eject much of their enriched gas into space through supernova explosions, this strongly suggests that stars are constantly forming out of interstellar gas – an idea we invoked in the last chapter to explain why galaxies do not all rapidly become red and dead.

Most galaxies contain huge amounts of gas and dust – here 'dust' means small solid particles – in very diffuse form between their

13. The 'Fingers of Creation'. Stars form within these columns of gas and dust, which are up to seven light years long

stars. The radiation spectra of these vast clouds of matter shows that they are very cool, sometimes only a few degrees Kelvin above absolute zero, and contain not only hydrogen, helium, and other heavy elements, but often molecules too. These giant molecular clouds are the obvious place for starbirth (Figure 13). For stars to form from them, bits of them must get much denser, presumably because they shrink under their own gravity. But like the clouds of water vapour in our own atmosphere, these clouds are not fixed entities always consisting of the same matter, but constantly losing and gaining it from other hotter phases of the interstellar medium such as supernovae and stellar winds. They resemble steam from a kettle, rather than vast self-gravitating collections of gas. Of course, gravity does act within them, but for the most part, the gas motions are rapid enough that there is no tendency for them to fall in on themselves.

Birth in obscurity

But purely by chance, *parts* of these clouds do feel self-gravity, or are compressed by some external event such as a supernova. Even then they may not collapse, because even this cold molecular gas can exert pressure. The ability of a clump of gas to resist gravity depends on whether it can raise its internal pressure fast enough to prevent infall. Now we know that pressure changes travel through a gas at the local speed of sound. If the gas clump is small, sound travels across it much faster than it can collapse under its own weight, so it stabilizes and oscillates, but does not collapse. However, if we consider larger and larger clumps, we eventually reach a size when the sound waves cannot travel through it and raise its pressure before it has already fallen in on itself and become much denser. This critical size depends on the ratio of temperature to mass density in the clump, and is called the Jeans length, after the British astronomer who first studied this problem in the early 20th century. The total mass of gas and dust inside a sphere with this radius is called the Jeans mass. A self-gravitating clump with a mass larger than this must start to collapse. In the Milky Way, this critical mass is thousands of solar masses.

Now things start to get complicated, because the very act of collapsing means that the density increases. The Jeans length and mass become smaller as a result, and individual parts of the originally collapsing clump now want to collapse separately – the clump fragments into smaller and smaller pieces. Fragmentation stops when the individual lumps become opaque to their own radiation. Instead of escaping freely, the trapped photons bounce around inside the clump of gas and force its internal temperature and pressure to rise, eventually to the point when they can balance gravity. This now prevents smaller parts of the lump of gas collapsing, and so stops further fragmentation. These clumps are still not yet stars: their only source of energy is the heat generated by the gentle contraction under gravity.

We considered a situation like this before, when in Chapter 2 we asked how the Sun can maintain itself against gravity. We found that the virial theorem tells the star that it must go on heating up internally as it loses energy by radiating it away from its surface. The timescale for this process is the thermal, or Kelvin–Helmholtz, time. This is about thirty million years for a clump with the mass of the Sun, but much shorter for massive stars, and much longer for low-mass stars. Eventually, after this thermal-timescale contraction, the centre of the contracting clump becomes hot enough and dense enough to start hydrogen burning, and the stars gradually join the main sequence. Finally, our star-formation event has given us a cluster of stars.

Astronomers try to predict the range of masses of the fragments, and how many objects there are at each mass, and compare them with the mass distribution observed in known star clusters. This distribution is called the initial mass function, or IMF. Its precise shape has a determining effect on how galaxies evolve. Observation suggests that in the real IMF, most stars have quite low masses, around the Sun's mass or less, with an ever-decreasing tail of stars with larger masses. It is likely that some of the fragments have masses too low ever to start nuclear burning, and form brown dwarfs. While some brown dwarfs are found by observations, they are so faint that we do not know how many form in a real cluster. And we also have little idea of how much of the gas in the collapsing cloud actually ends up as stars or brown dwarfs. The uncertainty here is that the radiation of each newly formed star drives off the dusty gas around it, preventing it from collapsing. We also do not understand well how binary stars form, and what the initial distribution of mass ratios is – do stars prefer companions of about their own size, or are they indifferent? These questions are urgent, as observationally we know that most massive stars are members of binary systems.

Theorists are actively working on these problems, using vast computer simulations. These simulations model the movements of gas like this mainly in one of two ways. One is to represent the theoretical gas as a collection of fictitious particles, whose movements under the forces of gravity and pressure the computer follows. The other way is to imagine that the space the gas occupies is divided into a grid of much smaller cells, and solving the equations relating gravity and pressure to motion to find how gas crosses from one cell to another. These methods are of course approximate, but give a very accurate idea of what really happens, provided that there are large numbers of fictitious particles or of grid cells covering the theoretical gas. This is directly analogous to the way a computer screen uses pixels to represent images. A million pixels would give a super-smooth picture of your face, and let you look closer and closer without seeing any blurring. But if there are only ten pixels, the image of your face will be blurred to the point that you are unrecognizable.

Computer simulations are just the same: the more particles or grid cells you can use, the more accurate the simulation, and the more you can believe the results. But the more particles or grid cells used to represent the gas motions, the longer the computer takes to work out what happens. So inevitably, the number of particles or grid cells that a simulation follows has to be kept within bounds. Big simulations now can use ten or even a hundred million particles, and similar numbers of grid cells, and may take weeks to follow the collapse into stars. The main difficulty is that the simulations must follow huge ranges of length and mass scales. The end product of the collapse of a large cloud of several parsecs size includes many stars smaller than the Sun, a factor of one hundred thousand times less. So what may have seemed like a large number of grid cells at the beginning of the simulation can become far too small to follow interactions between newly forming stars, since they would occupy only a few cells.

So once the action we are interested in happens on lengthscales which are much smaller than with those assumed at the start of the calculation, the computations cannot reliably follow what happens, because the details have become blurred. Putting in more information (more cells or particles) at the start of the calculations so that we can follow them further (technically, increasing the resolution of the simulations) means that they run prohibitively slowly. Ultimately of course, these problems will be overcome by increases in computer power. But for the moment, it is still difficult to be definite about how binary stars form – for example: do the two stars form as nearby fragments, or does a single star capture another gravitationally? If capture is common, this might make massive stars, simply by merging together two lower-mass fragments.

All this tells us that there is no hope of simulating the collapse of a cloud all the way down to individual stars. We have to make guesses about what is likely to happen at smaller lengthscales, and start our simulations there. For example, the general picture that stars form in clusters rather than in isolation strongly suggests that individual forming stars (protostars) must be built up from gas that has a lot of rotation – after all, the stars in a cluster move around under their mutual gravity, so the gas must be moving too. So we expect a protostar to be surrounded by an extended flattened disc of gas and dust, very similar to the accretion discs we met earlier in discussing X-ray binaries. This is is an obvious place for planets to form. But astronomers did not know of a single planet orbiting any normal star other than the Sun until Michel Mayor and Didier Queloz discovered the first in 1995. We now know of hundreds of planets orbiting nearby stars, and the observed sample is growing rapidly. Theorists are hard at work modelling these protostellar discs, and trying to predict the expected properties of the planets they produce, such as their masses and the sizes of their orbits around their parent star.

We know much less about how stars form than we do about any later part of their evolution. This has not stopped astronomers trying to work out even more ambitious things. One important question is what the first ever stars were like. These must have formed when there were no elements beyond hydrogen and helium – what were their masses, for example? The first galaxies must have been forming when the Universe was much smaller than it is now, so that collisions and mergers between them were probably common. This may well be how the galaxies we see now were formed. To understand this, we need to know what happens when galaxies collide, and in particular how many stars form, and how much gas may be blown out into space.

The last stars

The cyclic nature of star formation, with stars being born from matter chemically enriched by earlier generations, and expelling still more processed material into space as they die, defines a cosmic epoch – the epoch of stars. The end of this epoch will arrive only when the stars have turned all the normal matter of the Universe into iron, and left it locked in dead remnants such as black holes. Even then, cosmic evolution will not stop, for as

Stephen Hawking showed, quantum mechanics requires black holes to radiate as thermal bodies, with fantastically low temperatures. Once the Universe expands to the point when the background radiation has cooled even below this, these holes will begin, incredibly slowly, to radiate themselves away over aeons. Fortunately, this ultimately dreary endgame is still far from view – we live in a young Universe, only about twice as old as the Sun, and only slightly older than the oldest stars. This is no coincidence, of course, for intelligent life can only flourish in such an epoch, and must die out as the last stars fade.

We have come a long way, from asking how the Sun could survive, to trying to work out how galaxies like our own came to exist, and speculating on an unimaginably distant future. On the way, we have learned what makes the stars shine, how the chemical elements in our own bodies were made in stars, and how black holes form. And all this has come from applying the laws of physics to the simplest possible picture of a star as a ball of gas. Stellar evolution and its consequences are the grandest expression of the power of the laws of physics.

Further reading

Books going beyond the treatment of stellar structure and evolution given here inevitably involve considerably more mathematics. A good textbook at this level is *An Introduction to the Theory of Stellar Structure and Evolution*, by Dina Prialnik (Cambridge University Press, 2000).

For a fascinating insight into how the theory developed in the early 20th century, see *The Internal Constitution of the Stars*, by Arthur S. Eddington (Cambridge University Press, first published 1926, digitally reprinted 1999).

For more on black holes, accretion, and related topics, at a non-mathematical level, see *Gravity's Fatal Attraction*, by Mitchell Begelman and Martin Rees (Cambridge University Press, 2nd edn., 2010).

For a specialist mathematical treatment of these topics, see *Accretion Power in Astrophysics*, by Juhan Frank, Andrew King, and Derek Raine (Cambridge University Press, 3rd edn., 2002).

Index

A

B

C

D

N

P

Q

R

S

T

U

V

W

X

Z

COSMOLOGY

A Very Short Introduction

Peter Coles

What happened in the Big Bang? How did galaxies form? Is the universe accelerating? What is 'dark matter'? What caused the ripples in the cosmic microwave background?

These are just some of the questions today's cosmologists are trying to answer. This book is an accesible and non-technical introduction to the history of cosmology and the latest developments in the field. It is the ideal starting point for anyone curious about the universe and how it began.

'A delightful and accesible introduction to modern cosmology'

Professor J. Silk, Oxford University

'a fast track through the history of our endlessly fascinating Universe, from then to now'

J. D. Barrow, Cambridge University

www.oup.com/vsi

GALAXIES

A Very Short Introduction

John Gribbin

Galaxies are the building blocks of the Universe: standing like islands in space, each is made up of many hundreds of millions of stars in which the chemical elements are made, around which planets form, and where on at least one of those planets intelligent life has emerged. In this *Very Short Introduction*, renowned science writer John Gribbin describes the extraordinary things that astronomers are learning about galaxies, and explains how this can shed light on the origins and structure of the Universe.

www.oup.com/vsi

ONLINE CATALOGUE

A Very Short Introduction

Our online catalogue is designed to make it easy to find your ideal Very Short Introduction. View the entire collection by subject area, watch author videos, read sample chapters, and download reading guides.

http://fds.oup.com/www.oup.co.uk/general/vsi/index.html

SOCIAL MEDIA
Very Short Introduction

Join our community

www.oup.com/vsi

- Join us online at the official Very Short Introductions **Facebook** page.
- Access the thoughts and musings of our authors with our online **blog**.
- Sign up for our monthly **e-newsletter** to receive information on all new titles publishing that month.
- Browse the full range of Very Short Introductions online.
- Read **extracts** from the Introductions for free.
- Visit our library of **Reading Guides**. These guides, written by our expert authors will help you to question again, why you think what you think.
- If you are a teacher or lecturer you can order inspection copies quickly and simply via our website.